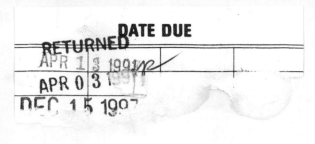

D1622875

Expanding Protons: Scattering at High Energies

Expanding Protons: Scattering at High Energies

Hung Cheng and Tai Tsun Wu

The MIT Press
Cambridge, Massachusetts
London, England

This book was set in Times Roman by Asco Trade Typesetting Ltd., Hong Kong, and printed and bound by Halliday Lithography in the United States of America.

Library of Congress Cataloging-in-Publication Data

Ch'eng, Hung.
  Expanding protons.

  Includes bibliographies and index.
  1. Protons—Scattering.   2. Nuclear reactions.   3. Scattering (Physics)
I. Wu, Tai Tsun.   II. Title.
QC793.5.P728C48   1987   539.7'54   86-27385
ISBN 0-262-03126-4

To Wen-Foo and Suh-Jen Yau and Paul C. Martin

Contents

Preface

This book covers a specialized field in elementary particle physics: the scattering and production of hadrons in the kinematic region of high energies and moderate transverse momentum transfers. Such processes are important because they are surprisingly rich in structure and furthermore because the corresponding cross sections are the only ones that are large and remain roughly constant with increasing center-of-mass energies.

Intensive studies of the high-energy scattering of hadrons began in the late 1950s, as accelerators capable of producing protons in the GeV region were built. By the early 1960s there emerged two prominent models: the Regge pole model and the droplet model of Yang and collaborators, both originally motivated by potential scattering. With the desire to go beyond potential scattering and to gain understanding of the various measured cross sections provided by experimentalists, theoretical physicists turned to quantum field theories for answers. Because of their simplicity, scalar field theories were the first ones explored, and the presence of Regge poles and Regge cuts was soon confirmed. It took several more years before it was realized that gauge field theories provide a more useful "laboratory" to test high-energy scattering theories. Indeed, the study of the high-energy behavior of gauge field theories has verified some of the important features of the droplet model and of the Regge pole model. More important, although the physical pictures of these two models differ from each other, they find a resolution in the theory that has emerged from this study of the high-energy behavior of gauge field theories. This theory is that of expanding protons.

One of the most surprising predictions of this theory is that, not only does the total cross section of hadron-hadron scattering not vanish as the center-of-mass energy increases without bound (as often conjectured in the 1960s), but it also rises indefinitely with energy. This means that the size of a proton expands indefinitely. Such an expansion was first observed in experiments performed at the Intersecting Storage Rings of CERN in 1973, three years after the theoretical prediction. A second surprising prediction of this theory is that the ratio of the integrated elastic cross section to the total cross section also increases, to a limiting value of $\frac{1}{2}$, as the center-of-mass energy increases without bound. Such an increase was first observed at the CERN $p\bar{p}$ collider in 1979, nine years after the theoretical prediction.

Because research in high-energy gauge field theories was extensive in the 1960s, the number of papers written on this topic is vast. Furthermore,

some of these papers are filled with complicated equations and demonstra-
tions of long calculations. This makes it difficult for researchers or graduate
students to find the most useful sources for materials needed and, even after
finding them, to digest and understand the contents. Consequently there
exists a general lack of communication in the field. It is for this reason that
there is a need for a book that gives a systematic exposition of this subject.
Our first draft of this book was indeed completed more than a decade ago.
But no sooner had we finished that we felt disenchanted with the work. We
believe that the complicated computations presented in the first draft mask
the beauty and the simplicity lying at the heart of the theory. It was not
until much later that we started anew, discarding almost entirely the first
draft. Our goal is to make the book comprehensible to a graduate student
in physics without watering down the contents. We have found to our grati-
fication that this is possible. Indeed, a presentation that emphasizes the
central idea rather than the details actually evokes deeper understanding
of the theory. The writing of this book was completed in early 1985, and
the literature search was carried out in 1984.

The first eight chapters of this book cover the main features of high-
energy scattering without going into any complicated calculations. Therefore it makes easy reading even for a graduate student who is taking a
course in quantum field theories. Indeed, chapter 1 gives a general survey
of the fundamentals of particle physics, and chapter 2 gives a self-contained
and succinct exposition of gauge field theories. We begin the study of
high-energy gauge field theories in chapter 3, where fermion-fermion elastic
scattering is studied. In chapter 4 we give a summary of the Regge pole
model and the droplet model. In chapters 5 and 6 we discuss inelastic
scattering in high-energy gauge field theories. By incorporating the results
of both elastic and inelastic scattering, we discuss in chapter 7 how they
lead to the theory of expanding protons. In chapter 8 we make the phenomenological analysis of experimental data. This chapter is perhaps the
one most useful for experimentalists or for people who wish to make phenomenological fits.

The last five chapters of the book are written for those researchers or
students who wish to learn how to carry out asymptotic calculations of
high-energy amplitude. We emphasize here that such calculations have the
undeserved reputation of being difficult. In truth, they are simple and

straightforward. In chapter 9 we extend the study in chapter 3 for pointlike fermions to particles with a structure. In chapter 10 we explain the method of integration in the momentum variables and show how easily high-energy amplitudes can be computed. In chapter 11 we apply the calculational method given in chapter 10 to high-energy scattering amplitudes of arbitrary order in Abelian gauge field theories. In chapter 12 we extend the study in chapter 11 to non-Abelian gauge field theories. Chapter 13 is the grand finale, where the Regge pole model and the droplet model find a resolution in the theory of expanding protons, which combines the crossed-channel unitarization and the direct-channel unitarization. The physical picture that a particle becomes more and more absorptive as energy increases is derived once again with more rigor. There are, in addition, a number of appendixes. In particular, appendix C gives a detailed discussion of the method of Feynman parameters and Mellin transform, which provides a way to compute high-energy amplitudes. This is an alternative to the method of momentum variables given in chapter 10.

A deep sense of gratitude is due Chen Ning Yang, who has given us guidance, shared his deep physical insights, been most generous with his time and attention, and provided the encouragement and inspiration that carried us through a most trying time. We have also benefited from working with C. Bourrely, E. Cantoral, and J. Soffer. We have enjoyed discussions with G. Bellettini, J. D. Bjorken, R. Cool, W. Dittrich, M. Gell-Mann, M. Goldberger, J. Jarlskog, F. Low, A. H. Mueller, C. A. Nelson, C. A. Orzalesi, and V. F. Weisskopf. We give our special thanks to L. Howard, who helped us with our numerical calculations. We are deeply indebted to R. W. P. King and C. C. Lin, who have exerted most important positive influence on our careers.

We are also very grateful to Margaret Owens, who expertly typed part of the manuscript, read the proof of the entire volume, and prepared both the author and subject indexes.

Our research on the expanding proton has been supported generously by the US Department of Energy (including its predecessors, the US Atomic Energy Commission and the US Energy Research and Development Administration), the National Science Foundation, and the Joint Services Electronics Program. The award of Guggenheim Fellowships in 1971 enabled us to write part of the first draft of this book, as well as to carry on further research on high-energy scattering. Finally, we thank CERN in Geneva, Switzerland, DESY in Hamburg, West Germany, and Fermilab in Illinois for their hospitality during our stays in their laboratories. All this support is gratefully acknowledged.

Hung Cheng
Massachusetts Institute of Technology
Tai Tsun Wu
Harvard University
April 1987

Expanding Protons: Scattering at High Energies

1 Introduction

Life is finite, whereas knowledge is infinite. It is difficult to pursue the infinite with the finite.

Chuang Tzu

1.1 Basic Constituents of Matter

Since the time of Democritus, physicists have been concerned with understanding the laws that deal with the behavior of the basic constituents in nature. Through the centuries our conception of what these constituents are has undergone many revolutions. A hundred years ago electricity, fluids, and other ponderable bodies were generally considered to be continuous, and whether or not they are infinitely divisible was discussed mostly on philosophical grounds. In the past century this concept has undergone three fundamental changes.

The first fundamental advance occurred in 1911, when Rutherford [1] inferred on the basis of the non-Gaussian distribution of the scattered particles from thin foils of matter that atoms have heavy nuclei. This discovery gave birth to nuclear physics. Soon the number of different nuclei increased rapidly, first one for each chemical element, then to many hundreds of isotopes [2].

That there are several hundred different basic constituents in nature is clearly unpalatable. Attempts to form nuclei from protons and electrons, the only available building blocks then besides the photon, were found to be impossible [3]. This dilemma was finally solved by the discovery of the neutron in 1932 by Chadwick [4] and by Curie-Joliot and Joliot [5]. Thus the second fundamental advance is the reduction of hundreds of isotopes to only two basic constituents: the proton and the neutron. Life becomes once more simple and satisfactory. Beginning in the 1950s, however, history repeated itself, and the number of these so-called strongly interacting particles, or hadrons, started to grow to well over a hundred [6].

Again, the presence of that many different basic constituents is unacceptable. Thus in 1964 Gell-Mann [7] and Zweig [8] proposed that hadrons are comprised of quarks. At that time it was not clear whether quarks were physical objects or merely mathematical entities. Ten years later, in 1975, observations at SPEAR of the Stanford Linear Accelerator Center established the physical presence of quarks in the form of jets [9]. Four years later, in 1979, gluons, which are responsible for the Yang-Mills interaction [10, 11] of quarks, were observed at PETRA of the Deutsches Elektronen-

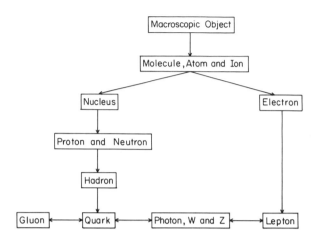

Figure 1.1
Structure of matter.

Synchrotron [12, 13]. Therefore, at present, we are in the middle of the third fundamental advance in the understanding of the basic constituents of matter: quarks, gluons, leptons, the photon, and the recently observed intermediate vector particles $W^{\pm}$ and Z [14, 15]. These advances are depicted schematically in figure 1.1.

Each fundamental advance brings with it fundamentally new problems. For example, a long time ago, the question of whether or not the neutron is basic was raised on the basis of its being unstable. Such questions of stability are now considered to be of no consequence; even the proton is sometimes believed to be unstable [16]. Today, a far more important question is whether quarks and gluons can be observed more directly than through the jets. Efforts to see fractionally charged quarks have not been generally successful, and accordingly there is a spectrum of opinions: Quarks can never be observed directly [17]; they are pair produced with very small cross sections; or they have some unusual property that makes them difficult to detect [18]. Which possibility is correct can be answered only experimentally.

1.2   High-Energy Scattering

The term *elementary particle* is not well defined. Roughly, an elementary particle is a particle that is smaller than an atom and is not a nucleus (except the hydrogen nucleus). According to this definition, an elementary particle may be a hadron, a lepton, a photon, a quark, a gluon, an intermediate boson W or Z, or perhaps one of the particles yet to be observed.

The determination of a hadron's or a lepton's electric charge and mass is usually fairly straightforward. Although the electric charge of a quark can be obtained from the cross section for $e^+e^- \to q\bar{q}$, even the masses are not experimentally known directly for quarks and gluons.

How do we proceed to learn more about these elementary particles? Because the laws that govern the behavior of elementary particles are revealed when the particles interact with one another, one of the central efforts in physics now is to establish these laws through particle scattering or annililation. Historically, one of the most successful scattering experiments is that of Rutherford [1].

The general principles of quantum mechanics and special relativity have produced many successful predictions and are firmly established. Therefore, whatever the laws are for elementary-particle physics, they are believed to satisfy these principles. The detailed nature of interaction, beyond these general constraints, is referred to as *dynamics* and is extremely rich and interesting. Because the manifestations of dynamics are undoubtedly complicated, we must look for situations in which some form of simplification prevails.

There are several distinct circumstances under which great simplifications are expected to occur and hence theoretical understanding is furthered. The best known circumstances are those for which a small parameter of expansion is available for perturbation treatments. Two well-known cases are the electromagnetic and weak interactions, at least at energies that are not too high. An example of such success is the astonishing agreement between the experimental results [19] and the theoretical computation of the Lamb shift [20].

There is a second circumstance under which simplifications occur: the symmetry properties of the interaction. Such symmetry considerations have played a most important role in the development of elementary-particle physics, especially during the last thirty years. Many examples are available, and here let us mention only one: the standard model of unified electroweak interactions of Glashow, Weinberg, and Salam [14].

We are concerned here mainly with yet another circumstance. This is the case of extremely high energies, where the wavelength of the incident particle is much smaller than the sizes of the interacting particles. Because it is perhaps much less obvious than in the previous two cases why there is any simplification at high energies, let us consider an analog in some detail. When a macroscopic object is illuminated by an electromagnetic wave, the resulting field is, in general, fairly complicated and not easily calculated. If the wavelength of the illuminating electromagnetic wave is small compared with the dimensions of the object, however, geometrical optics takes over, and reflection and refraction become a suitable and sufficient description for many situations. As a concrete example, let us imagine that we are in a room equipped with a red light bulb at one instance and with a blue light bulb at a later time. It is a common experience that, as the light bulb is changed, the objects in the room change their colors but *not* their shapes and sizes. This implies, in particular, that the scattering cross section does not depend on the light frequency. That the shapes and sizes are independent of the wavelength simplifies the physics—and our daily life—greatly. Just imagine trying to sit in a chair that has different shapes for different color rays in the sunbeam!

On the basis of this optical analog, it is perhaps not unreasonable to expect simplification in the physics of elementary particles when the frequency, and hence the energy, is extremely high. For definiteness, let us consider the scattering of a proton by another proton. Because the mass of the proton (multiplied by c^2) is approximately 1 GeV (or 10^9 eV), roughly the energy may be considered to be high if each proton has an energy of 10 GeV in the center-of-mass system. By special relativity, this corresponds to shooting at a stationary target with 200-GeV protons. This is typically considered to be the beginning of the high-energy region of interest.

1.3   Relativistic Quantum Field Theory

What procedure can we follow to study strong interactions at extremely high energies? Before attempting to answer this question, let us review briefly our knowledge of electromagnetic and weak interactions. In the late 1920s Dirac, Fermi, Heisenberg, Jordan, and Pauli [21] united quantum mechanics with electromagnetic theory, giving birth to quantum electrodynamics. This was accomplished by replacing the field quantities in classical Maxwell theory by field operators, which are functions of space and time and satisfy appropriate commutation relations. Field theories that embody both special relativity and quantum mechanics are known as *relativistic quantum field theories*, and quantum electrodynamics is the first such example. The first half of the 1930s was a period of great activity, as numerous scattering processes were understood on the basis of quantum electrodynamics. The Klein-Nishina formula was found for Compton scattering (photon-electron scattering); the Bethe-Heitler formula for bremsstrahlung (radiation from an electron in a field) and the formulas for Møller scattering (electron-electron scattering), Bhabha scattering (electron-positron scattering), pair production, and Delbrück scattering (scattering of a photon by a static Coulomb field) were all calculated [22]. At about the same time, Fermi also invented his celebrated relativistic quantum field theory for weak interactions [23].

The ensuing development of the theory of weak interactions took forty years, the milestones of this journey being the introduction of parity non-conservation by Lee and Yang [24] and the aforementioned unification of weak interaction and electromagnetic interaction by Glashow, Weinberg, and Salam [14]. Today it is universally accepted that relativistic quantum field theory accounts correctly for both electromagnetic interaction and weak interaction.

The field theories for the electromagnetic and weak interactions are of a special kind: gauge field theories [10]. It is believed that the field theory underlying the interaction of hadrons is also a gauge field theory. This is discussed in more detail in the next chapter. Our procedure of studying high-energy hadron scattering is therefore to investigate the scattering amplitude in quantum gauge theory.

Unlike the coupling constants in electroweak interactions, which are small, the coupling constant in the strong interaction of hadrons is of order 1, at least for small momentum transfers [25]. Therefore, to study strong interactions, one should not calculate only the low-order terms by perturbation theory; instead it is necessary to study systematically perturbation theory to all orders.

Although, in principle, the high-energy behavior of a gauge field theory can depend qualitatively on the details of the coupling, this in fact is not the case. Instead, *general features* are common to all gauge field theories. The program therefore involves the extraction of these general features and then the application of these features to hadron scattering at high energies and small momentum transfers. To avoid unnecessary complications that merely obscure the main issues, we have decided to treat in detail the simplest case of the Abelian gauge theory and to defer treatment of the more complicated Yang-Mills non-Abelian case to chapters 12 and 13.

References

[1] E. Rutherford, *Philosophical Magazine* 21 (1911):669.

[2] J. J. Thomson, *Rays of Positive Electricity* (London: Longmans, Green, 1913).

[3] P. Ehrenfest and J. R. Oppenheimer, *Physical Review* 37 (1931):333.

[4] J. Chadwick, *Proceedings of the Royal Society of London*, ser. A, 136 (1932):692.

[5] I. Curie-Joliot and F. Joliot, *Comptes Rendus*, Academy of Science (Paris), 194 (1932):273.

[6] See, for example, any issue of *Review of Particle Properties* by the Particle Data Group. The most recent review is in *Reviews of Modern Physics* 58 (1986):S1.

[7] M. Gell-Mann, *Physics Letters* 8 (1964):214.

[8] G. Zweig, CERN Reports TH-401, TH-412 (1964).

[9] G. Hanson, G. S. Abrams, A. M. Boyarski, M. Breidenbach, F. Bulos, W. Chinowsky, G. J. Feldman, C. E. Friedberg, D. Fryberger, G. Goldhaber, D. L. Hartill, B. Jean-Marie, J. A. Kadyk, R. R. Larsen, A. M. Litke, D. Lüke, B. A. Lulu, V. Lüth, H. L. Lynch, C. C. Morehouse, J. M. Paterson, M. L. Perl, F. M. Pierre, T. P. Pun, P. A. Rapidis, B. Richter, B. Sadoulet, R. F. Schwitters, W. Tanenbaum, G. H. Trilling, F. Vannucci, J. S. Whitaker, F. C. Winkelmann, and J. E. Wiss, *Physical Review Letters* 35 (1975):1609.

[10] C. N. Yang and R. L. Mills, *Physical Review* 96 (1954):191.

[11] J. Ellis, M. K. Gaillard, and G. G. Ross, *Nuclear Physics B* 111 (1976):253 (errata, *Nuclear Physics B* 130 (1977):516); T. A. DeGrand, Y. J. Ng, and S. H. H. Tye, *Physical Review D* 16 (1977):3251; A. DeRujula, J. Ellis, E. G. Floratos, and M. K. Gaillard, *Nuclear Physics B* 138 (1978):387.

[12] S. L. Wu and G. Zobernig, TASSO Note 84 (1979); S. L. Wu and G. Zobernig, *Zeitschrift für Physik C* 2 (1979):107; B. H. Wiik, *Proceedings of the International Conference on Neutrinos, Weak Interactions, and Cosmology* (Bergen, Norway, 1979), 113; P. Söding, *Proceedings of the European Physics Society International Conference on High Energy Physics* (Geneva, Switzerland, 1979), 271; TASSO Collaboration (R. Brandelik, W. Braunschweig, K. Gather, et al.), *Physics Letters B* 86 (1979):243; MARK J Collaboration (D. P. Barber, U. Becker, H. Benda, et al.) *Physical Review Letters* 43 (1979):830; PLUTO Collaboration (C. Berger, H. Genzel, R. Grigull, et al.), *Physics Letters B* 86 (1979):418; JADE Collaboration (W. Bartel, T. Canzler, D. Cords, et al.), *Physics Letters B* 91 (1980):142.

[13] S. L. Wu, *Physics Reports* 107 (1984):59.

[14] S. L. Glashow, *Nuclear Physics* 22 (1961):579; S. Weinberg, *Physical Review Letters* 19 (1967):1264; A. Salam, in *Eighth Nobel Symposium*, N. Svartholm, ed. (New York: Wiley, 1968), 367.

[15] UA1 Collaboration (G. Arnison, A. Astbury, B. Aubert, et al.) *Physics Letters B* 122 (1983):103, 126 (1983):398; UA2 Collaboration (P. Bagnaia, M. Banner, R. Battiston et al.), *Physics Letters B* 122 (1983):476, 129 (1983):130.

[16] H. Georgi and S. L. Glashow, *Physical Review Letters* 32 (1974):438; J. C. Pati and A. Salam, *Physical Review D* 10 (1974):275.

[17] J. Schwinger, *Physical Review* 125 (1962):397, 128 (1962):2425; J. H. Lowenstein and J. A. Swieca, *Annals of Physics* (New York) 68 (1971):172; M. Gell-Mann, *Acta Physica Austriaca*, suppl. 9 (1972):733; H. Fritzsch and M. Gell-Mann, *Proceedings of the Sixteenth International Conference on High Energy Physics* (Batavia, Ill.: National Accelerator Laboratory, 1972), vol. 2, 135; D. Amati and M. Testa, *Physics Letters B* 48 (1974):227; P. Olesen, *Physics Letters B* 50 (1974):255; A. Chodos, R. L. Jaffe, K. Johnson, C. B. Thorn, and V. F. Weisskopf, *Physical Review D* 9 (1974):3471; T. T. Wu, B. M. McCoy, and H. Cheng, *Physical*

Review D 9 (1974): 3495; J. Kogut and L. Susskind, *Physical Review D* 9 (1974): 3501; A. Casher, J. Kogut, and L. Susskind, *Physical Review D* 10 (1974): 732; K. G. Wilson, *Physical Review D* 10 (1974): 2445.

[18] B. M. McCoy and T. T. Wu, *Physics Letters B* 72 (1977): 219, *Physics Reports* 49 (1979): 193; G. Barbiellini, G. Bonneaud, R. J. Cashmore, G. Coignet, J. Ellis, M. K. Gaillard, J. F. Grivaz, C. Matteuzzi, R. D. Peccei, and B. H. Wiik, DESY Report 80-42 (1980).

[19] W. E. Lamb Jr. and R. C. Retherford, *Physical Review* 72 (1947): 241, 75 (1949): 1325, 79 (1950): 549, 81 (1951): 222, 85 (1952): 259, 86 (1952): 1014; S. Triebwasser, E. S. Dayhoff, and W. E. Lamb Jr., *Physical Review* 89 (1953): 98, 106.

[20] H. A. Bethe, *Physical Review* 72 (1947): 339; N. M. Kroll and W. E. Lamb Jr., *Physical Review* 75 (1949): 388; J. B. French and V. F. Weisskopf, *Physical Review* 75 (1949): 1240; R. P. Feynman, *Physical Review* 74 (1948): 1430, 76 (1949): 769 n13; J. Schwinger, *Physical Review* 76 (1949): 790; H. Fukuda, Y. Miyamoto, and S. Tomonago, *Progress of Theoretical Physics* 4 (1948): 47, 121.

[21] P. A. M. Dirac, *Proceedings of the Royal Society of London* 112 (1926): 661, 114 (1927): 243, 710; P. Jordan and W. Pauli, *Zeitschrift für Physik* 47 (1928): 151; E. Fermi, *Rendiconti della R. Accademia Nationale del Lincei* 9 (1929): 881, 12 (1930): 431; W. Heisenberg and W. Pauli, *Zeitschrift für Physik* 56 (1929): 1, 59 (1930): 168.

[22] O. Klein and Y. Nishina, *Zeitschrift für Physik* 52 (1929): 853; I. Tamm, *Zeitschrift für Physik* 62 (1930): 545; C. Møller, *Annalen der Physik* 14 (1932): 531; M. Delbrück, *Zeitschrift für Physik* 84 (1933): 144; H. A. Bethe and W. Heitler, *Proceedings of the Royal Society of London*, ser. A, 146 (1934): 83; H. J. Bhabha, *Proceedings of the Royal Society of London*, ser. A, 154 (1936): 195; G. Racah, *Nuovo Cimento* 13 (1936): 66.

[23] E. Fermi, *Zeitschrift für Physik* 88 (1934): 161.

[24] T. D. Lee and C. N. Yang, *Physical Review* 104 (1956): 254.

[25] G. 't Hooft and M. Veltman, *Renormalization of Yang-Mills Fields and Applications to Particle Physics*, C. P. Korthals-Altes, ed. (Marseilles: CNRS, 1972); H. D. Politzer, *Physical Review Letters* 30 (1973): 1346; D. J. Gross and F. Wilczek, *Physical Review Letters* 30 (1973): 1343.

2 Gauge Field Theory

The colors are no more than five in number, but their mixtures are so numerous that one cannot visualize them all.

The flavors are no more than five in number, but their blends are so various that one cannot taste them all.

In battle there are only the Chi (strange) and Cheng (normal) forces, but their combinations are limitless; none can comprehend them all.

Sun Tzu on War

2.1 Basic Principles

In the development of quantum mechanics the concept of measurement, introduced and developed by Bohr [1] plays a central role. For example, the Heisenberg uncertainty principle [2] must be understood in terms of the various ways of measuring the position and momentum of a particle. Thus the fundamental quantities in quantum mechanics are the observables.

Gauge field theory is based on a strengthened version of this basic principle: Only physically measurable quantities can be given by a physical theory [3].

Let us illustrate this basic principle with the simplest example. In non-relativistic quantum mechanics the wave function of a particle is, in general, a complex function of space and time. With suitable normalization, the absolute value squared of the wave function $\Phi(\mathbf{x}, t)$ is the probability of observing the particle at $\mathbf{x}$ and t. On the other hand, the phase of Φ does not have such a direct physical interpretation. More precisely, although the phase difference between two wave functions leads to interference, multiplying all wave functions by the same phase factor has no measurable effects. Thus it is natural to declare that the two representations of wave functions related by

$$\Phi'(\mathbf{x}, t) = e^{i\alpha}\Phi(\mathbf{x}, t) \tag{2.1}$$

are equivalent, where α is a real constant. It then follows that Φ and Φ' should satisfy the same Schrödinger equation. Indeed, it is easy to verify that this is the case, as we can eliminate the common phase factor from the equation. The transformation (2.1) between Φ and Φ' with α constant is called a gauge transformation of the first kind.

What happens if α is a function of space and time? In this case

$$\frac{\partial}{\partial t}\Phi = e^{-i\alpha}\left(\frac{\partial}{\partial t} - i\frac{\partial \alpha}{\partial t}\right)\Phi' \tag{2.2}$$

and

$$\nabla\Phi = e^{-i\alpha}(\nabla - i\nabla\alpha)\Phi'. \tag{2.3}$$

Because of the terms involving the derivatives of α in equations (2.2) and (2.3), the equation satisfied by Φ may be different from that satisfied by Φ'. The transformation (2.1), with α a function of space and time, is called the gauge transformation of the second kind.

If we insist that the equation satisfied by the wave function be invariant under gauge transformations of the second kind, a profound result is obtained. Let us recall that in the Schrödinger equation [4] with a potential V, the time derivative and the potential always appear in the combination of $(i(\partial/\partial t) - eV)$. Also, in the presence of a vector potential $\mathbf{A}$, the gradient and $\mathbf{A}$ always appear in the combination of $(-i\nabla - e\mathbf{A})$. Now we have from equations (2.2) and (2.3) that

$$\left(\frac{\partial}{\partial t} + ieV\right)\Phi = e^{-i\alpha}\left(\frac{\partial}{\partial t} + ieV'\right)\Phi' \tag{2.4}$$

and

$$(\nabla - ie\mathbf{A})\Phi = e^{-i\alpha}(\nabla - ie\mathbf{A}')\Phi', \tag{2.5}$$

where

$$V' = V - \frac{1}{e}\frac{\partial\alpha}{\partial t} \tag{2.6}$$

and

$$\mathbf{A}' = \mathbf{A} + \frac{1}{e}\nabla\alpha. \tag{2.7}$$

Thus the Schrödinger equation for Φ' is of the same form as that for Φ, with $(V', \mathbf{A}')$ taking the place of $(V, \mathbf{A})$. We observe that the transformation between $(V', \mathbf{A}')$ and $(V, \mathbf{A})$ given by equations (2.6) and (2.7) is precisely the gauge transformation [5] in electromagnetism, with $(V', \mathbf{A}')$ and $(V, \mathbf{A})$ corresponding to exactly the same electric field $\mathbf{E}$ and magnetic field $\mathbf{B}$. We can therefore say that gauge invariance of the second kind is preserved by the existence of the electromagnetic field. A more dramatic view is that the invariance under gauge transformation of the second kind begs the existence of the electromagnetic field—hence the motivation behind the command, "Let there be light."

2.2   Quantum Electrodynamics

The gauge transformation (2.1) holds for the wave function describing the
state of a single particle. In quantum field theories it is necessary to treat
states with many particles. Indeed, because particles can be created or
destroyed, a state need not contain a fixed number of particles. Therefore,
to formulate the gauge principle in quantum field theories, we must extend
the treatment presented in section 2.1.

  A transformation is generally represented by a unitary operator U that
transforms a state vector $|\Phi\rangle$ into $U|\Phi\rangle$:

$$|\Phi\rangle \to |\Phi'\rangle = U|\Phi\rangle. \tag{2.8}$$

Under this transformation the field operator ψ of an electron (as well as
all other operators) must transform according to

$$\psi \to \psi' = U\psi U^{-1} \tag{2.9}$$

so that

$$\langle \Phi_1 |\psi|\Phi_2\rangle = \langle \Phi_1'|\psi'|\Phi_2'\rangle. \tag{2.10}$$

Because ψ is the operator that either annihilates an electron or creates a
positron, the nonvanishing elements in equation (2.10) are those with the
charges of the two states differing by 1. To compensate for the difference
of the phase factors of these two states under gauge transformations, we
must have, to satisfy equation (2.10),

$$\psi \to \psi' = e^{i\alpha}\psi. \tag{2.11}$$

  If ψ' can serve just as well as ψ as the fermion field operator, then ψ' and
ψ should satisfy the same equation of motion. Because this equation is
derived from the Lagrangian, we must check whether or not the Lagrangian
is invariant under the transformation (2.11). The Lagrangian (or perhaps
more precisely, the Lagrangian density) for a free electron field is given by
[6]

$$\mathscr{L}_0 = \bar{\psi}(i\partial\!\!\!/ - m)\psi, \tag{2.12}$$

where m is the electron mass and (using the summation convention)

$$\partial\!\!\!/ \equiv \gamma^\mu \partial_\mu,$$

with γ^μ the 4×4 Dirac matrices, a representation of which is explicitly given in chapter 3. The Lagrangian $\mathscr{L}_0$ is invariant under gauge transformations of the first kind. As a consequence of this invariance, there exists, according to Noether's theorem [7], a conserved quantity that is the electric charge.

The Lagrangian $\mathscr{L}_0$ is not invariant under gauge transformations of the second kind and hence neither is the equation of motion for ψ. In exactly the same way as in the preceding section, we introduce a 4-vector field A^μ coupled to the fermion field in the combination of

$$\partial_\mu + ieA_\mu. \tag{2.13}$$

Then the Lagrangian for the fermion field becomes

$$\bar{\psi}(i\not\partial - e\not A - m)\psi, \tag{2.14}$$

where

$$\not A = \gamma^\mu A_\mu. \tag{2.15}$$

Under the gauge transformation (2.11), A_μ transforms according to equations (2.6) and (2.7), which can be summarized by

$$A_\mu \to A'^\mu = A^\mu - \frac{1}{e}\partial^\mu\alpha. \tag{2.16}$$

It then follows that the Lagrangian (2.14) is invariant under all gauge transformations. The operators A^μ are the operators for the electromagnetic field. Thus the requirement of invariance under gauge transformations of the second kind leads to the existence of photons.

By adding to expression (2.14) the term $-\frac{1}{4}F_{\mu\nu}F^{\mu\nu}$, the Lagrangian of the free electromagnetic field, we obtain the Lagrangian of quantum electrodynamics (QED) [8]:

$$\mathscr{L} = -\tfrac{1}{4}F_{\mu\nu}F^{\mu\nu} + \bar{\psi}(i\not\partial - e\not A - m)\psi, \tag{2.17}$$

where

$$F^{\mu\nu} = \partial^\mu A^\nu - \partial^\nu A^\mu. \tag{2.18}$$

Note that $F^{\mu\nu}$ is unchanged under gauge transformations. Thus $\mathscr{L}$ of equation (2.17) is invariant under all gauge transformations.

We can similarly derive the Lagrangian of scalar electrodynamics. The Lagrangian for a free complex scalar field ϕ of mass m is given by

$$\partial_\mu \phi^+ \partial^\mu \phi - m^2 \phi^+ \phi,$$

where ϕ^+ is the Hermitian conjugate of ϕ. By the gauge principle, we make the replacement

$$\partial^\mu \to \partial^\mu + ieA^\mu. \tag{2.19}$$

The Lagrangian of a charged scalar field coupled to an electromagnetic field is

$$[(\partial^\mu + ieA^\mu)\phi]^+ (\partial_\mu + ieA_\mu)\phi - m^2 \phi^+ \phi - \tfrac{1}{4} F_{\mu\nu} F^{\mu\nu}. \tag{2.20}$$

2.3   Non-Abelian Gauge Fields

In 1954, Yang and Mills [3] extended the gauge principle to hadron physics. This extension provides a powerful tool for the construction of quantum field theories relevant to particle physics. For the clarity of presentation and for historical perspectives, in this section let us follow closely the arguments of Yang and Mills.

The proton p and the neutron n have nearly equal masses. Furthermore, the n-p, p-p, and n-n forces are nearly equal, provided that the spin and the orbital angular momentum states are the same. It is believed that the slight difference in mass and force is due to the fact that the proton is charged, whereas the neutron is neutral, and that these two particles would have the same properties if electromagnetism were turned off. Recognizing this equality, Heisenberg [9] proposed in 1932 the description of p and n as two charged states of a nucleon. More precisely, in analogy with the spin of a fermion, a spin of $\tfrac{1}{2}$ in an abstract space is given to the nucleon. This space is called the space of the isotopic spin. (We refer to it as the I-space.) The proton corresponds to the state with $I_z = \tfrac{1}{2}$, whereas the neutron corresponds to the state with $I_z = -\tfrac{1}{2}$. The equality of the properties of the proton and the neutron is expressed by the invariance, in the absence of the electromagnetic interaction, of the Lagrangian under rotations in the I-space. In this way it makes no difference what we choose to call a proton and a neutron. The equations of motion remain the same.

Thus in the I-space the state of a nucleon is represented by a two-component wave function:

$$\psi = \begin{pmatrix} \psi_p \\ \psi_n \end{pmatrix}. \tag{2.21}$$

Under a rotation in the I-space, both the new proton and the new neutron states are linear superpositions of the old proton and neutron states. This is expressed by

$$\psi \rightarrow \psi' = U\psi, \tag{2.22}$$

where U is a 2×2 matrix. To preserve the normalization of ψ, we require U to be unitary. Thus

$$U = e^{i\alpha}, \tag{2.23}$$

where α is Hermitian. These 2×2 unitary matrices form a group known as $U(2)$. Comparing equation (2.23) with equation (2.1), we see that the only difference is that α in equation (2.23) is a 2×2 Hermitian matrix, whereas α in equation (2.1) is a real number.

There exist four independent 2×2 Hermitian matrices. We choose them to be the unity matrix I and the traceless matrices of Pauli, σ_a, $a = 1, 2,$ 3 ($\sigma_1 = \left[\begin{smallmatrix} 0 & 1 \\ 1 & 0 \end{smallmatrix}\right]$, $\sigma_2 = \left[\begin{smallmatrix} 0 & -i \\ i & 0 \end{smallmatrix}\right]$, $\sigma_3 = \left[\begin{smallmatrix} 1 & 0 \\ 0 & -1 \end{smallmatrix}\right]$). If $\alpha = \theta I$, where θ is a real number, then U is equal to $e^{i\theta}I$ and the transformation adds the same phase to the proton and the neutron but does not mix them. These complex phase factors form a group. Because all members of the group commute, the group is Abelian and is known as $U(1)$. The transformations of this group are exactly the same as those discussed in section 2.2. Invariance of the Lagrangian under this group leads to conservation of nucleon number in the same way that invariance of the Lagrangian under the gauge transformations of the electron wave function leads to conservation of electric charge. We do not discuss these transformations any further.

Next, let

$$\alpha = \theta_a T_a, \tag{2.24}$$

where

$$T_a = \tfrac{1}{2}\sigma_a \tag{2.25}$$

and the repeated indexes imply a summation over $a = 1, 2, 3$. Note that, because α is traceless, the determinant of U equals 1:

$$\det U = e^{i\,\mathrm{Tr}\,\alpha} = 1.$$

These 2×2 unitary matrices with determinant 1 form a group called $SU(2)$.

The infinitesimal transformations of $SU(2)$ are of the form

$$1 + i \sum_{i=1}^{3} \delta\theta_a T_a,$$

and the infinitesimal generators of this group are T_a, $a = 1, 2, 3$. They satisfy the Lie algebra

$$[T_a, T_b] = i\varepsilon_{abc} T_c, \tag{2.26}$$

where ε_{abc} is completely antisymmetric and $\varepsilon_{123} = 1$. Because the infinitesimal generators do not commute, the group $SU(2)$ is a non-Abelian group. The 2×2 matrices obey, in addition,

$$\text{Tr } T_a T_b = \tfrac{1}{2}\delta_{ab}, \tag{2.27}$$

$$[T_a, T_b]_+ = \tfrac{1}{2}\delta_{ab}, \tag{2.28}$$

where $[\ ]_+$ denotes the anticommutator. The Lie algebra (2.26) is exactly the same as that of the three-dimensional rotation group $O(3)$. Therefore, just as the invariance of a Lagrangian under rotation gives rise to conservation of angular momentum, the invariance of a Lagrangian under gauge transformations of $SU(2)$ gives rise to conservation of isotopic spin.

We have so far implicitly assumed that α is independent of space and time. Thus equation (2.23) is a gauge transformation of the first kind. We now require that the Lagrangian be invariant under gauge transformations of $SU(2)$ chosen independently at every space-time point. We are then naturally led to the concept of non-Abelian gauge fields.

We demonstrate this by essentially repeating the arguments of the preceding section. We have from equation (2.22) that

$$\partial_\mu \psi = \partial_\mu (U^{-1}\psi') = U^{-1}[\partial_\mu + U(\partial_\mu U^{-1})]\psi'. \tag{2.29}$$

The term involving $\partial_\mu U^{-1}$ in equation (2.29) may cause a change of the equation of motion; this can be avoided if we introduce a set of gauge fields. If U is unitary and of unit determinant, $iU(\partial_\mu U^{-1})$ is a Hermitian and traceless matrix, which has three independent components. (To show that $iU(\partial_\mu U^{-1})$ is Hermitian, we make use of the fact that, because $UU^{-1} = I$, we have

$$(\partial_\mu U)U^{-1} + U(\partial_\mu U^{-1}) = 0.$$

To show that $iU(\partial_\mu U^{-1})$ is traceless, we make use of the identity

$$\mathrm{Tr}(U^{-1}\partial_\mu U) = \frac{1}{\det U}\partial_\mu(\det U),$$

which can be proved by writing both sides of this identity in terms of the matrix elements of U and their cofactors.)

Thus we need to introduce three gauge fields A_a^μ, $a = 1, 2, 3$, which form the components of a Hermitian and traceless 2×2 matrix A^μ:

$$A^\mu \equiv \frac{1}{2}\begin{bmatrix} A_3^\mu & A_1^\mu - iA_2^\mu \\ A_1^\mu + iA_2^\mu & -A_3^\mu \end{bmatrix} = A_a^\mu T_a. \tag{2.30}$$

We then have, from equation (2.29), that

$$(\partial_\mu + igA_\mu)\psi = U^{-1}[\partial_\mu + U(\partial_\mu U^{-1}) + igUA_\mu U^{-1}]\psi'$$
$$= U^{-1}(\partial_\mu + igA_\mu')\psi', \tag{2.31}$$

where

$$A_\mu' = UA_\mu U^{-1} + \frac{1}{ig}U(\partial_\mu U^{-1}). \tag{2.32}$$

Note that, if A_μ is Hermitian and traceless, so is A_μ'. Note also that A_μ does not transform covariantly, that is,

$$A_\mu' \neq UA_\mu U^{-1}.$$

This is because A_μ is, in addition to being an isovector of three components, a gauge field and has to absorb the term

$$\frac{1}{ig}U(\partial_\mu U^{-1})$$

in a gauge transformation. It is convenient to introduce the notation

$$D_\mu \equiv \partial_\mu + igA_\mu. \tag{2.33}$$

By equation (2.31), we have

$$D_\mu = U^{-1}D_\mu'U. \tag{2.34}$$

D_μ is called the gauge covariant derivative. For an infinitesimal transformation

$$U = 1 + i\delta\theta,$$

where $\delta\theta$ is an infinitesimal matrix, equation (2.32) becomes

$$A'_\mu - A_\mu = i[\delta\theta, A_\mu] - \frac{1}{g}\partial_\mu\delta\theta$$

$$= -\frac{1}{g}[D_\mu, \delta\theta]. \tag{2.35}$$

Note the similarity with equation (2.16).

With the gauge covariant derivative it is easy to write a gauge invariant Lagrangian for the nucleon field. It is

$$\bar{\psi}(i\slashed{D} - M)\psi, \tag{2.36}$$

where M is the nucleon mass and ψ is now used to denote the two-component nucleon. Expression (2.36) is the counterpart of expression (2.14) for the Lagrangian of the electron field.

It remains to construct a gauge invariant Lagrangian for the gauge field. Again, drawing on an analogy with the electromagnetic field, we would like to find the counterpart of $F_{\mu\nu}$ defined in equation (2.18). This counterpart is required to be covariant under gauge transformations. It is straightforward to check that the naive guess does not work, that

$$\partial_\mu A_\nu - \partial_\nu A_\mu$$

does not transform covariantly under gauge transformation. Quite simply, neither ∂_μ nor A_ν is covariant. One of the simplest ways to construct covariant quantities is to make use of the fact that D_μ is covariant; therefore so is $[D_\mu, D_\nu]$. Because $[\partial_\mu, \partial_\nu] = 0$, we have

$$F_{\mu\nu} \equiv \frac{1}{ig}[D_\mu, D_\nu] = \partial_\mu A_\nu - \partial_\nu A_\mu + ig[A_\mu, A_\nu]. \tag{2.37}$$

Thus $F_{\mu\nu}$ is a 2×2 matrix that is gauge covariant:

$$F'_{\mu\nu} = UF_{\mu\nu}U^{-1}.$$

The counterpart of the Lagrangian $-\frac{1}{4}F_{\mu\nu}F^{\mu\nu}$ for the electromagnetic field is therefore the gauge invariant expression

$$-\frac{1}{2}\text{Tr}\,F_{\mu\nu}F^{\mu\nu} = -\frac{1}{4}F^a_{\mu\nu}F^{\mu\nu,a}, \tag{2.38}$$

where

$$F_{\mu\nu} = F^a_{\mu\nu}T^a. \tag{2.39}$$

A more explicit expression for $F^a_{\mu\nu}$ is obtained from equation (2.39) by utilizing equations (2.26) and (2.27):

$$F^a_{\mu\nu} = 2\,\text{Tr}(F_{\mu\nu}T_a)$$

$$= \partial_\mu A^a_\nu - \partial_\nu A^a_\mu - g\varepsilon_{abc}A^b_\mu A^c_\mu. \tag{2.40}$$

The complete Lagrangian for the nucleon field coupled to the gauge vector field is therefore given by

$$\bar\psi(i\slashed{D} - M)\psi \quad \tfrac{1}{4}F^a_{\mu\nu}F^{\mu\nu,a}. \tag{2.41}$$

Similarly, we can also derive the Lagrangians of other fields coupled to the A^a_μ fields. Consider, for example, a complex scalar field ϕ that is of isospin $\tfrac{1}{2}$. We express ϕ as

$$\phi = \begin{pmatrix} \phi_1 \\ \phi_2 \end{pmatrix}, \tag{2.42}$$

where ϕ_1 and ϕ_2 are complex scalars. (This is in contrast to equation (2.21), where ψ_p and ψ_n are Dirac spinors.) Under a gauge transformation,

$$\phi \to \phi' = U\phi.$$

With reference to expression (2.20), a gauge-invariant Lagrangian involving ϕ is

$$(D_\mu\phi)^+(D_\mu\phi) - \mu^2\phi^+\phi, \tag{2.43}$$

where μ is the mass of the meson.

Next consider a scalar field V that is of isospin 1. This field has three real components, V_1, V_2, and V_3. As in identity (2.30), we can express this field by a 2×2 Hermitian and traceless matrix:

$$V = \frac{1}{2}\begin{bmatrix} V_3 & V_1 - iV_2 \\ V_1 + iV_2 & -V_3 \end{bmatrix} = V_a T^a. \tag{2.44}$$

Under a gauge transformation,

$$V \to V' = UVU^{-1}. \tag{2.45}$$

A gauge-invariant Lagrangian involving V is

$$\tfrac{1}{2}\text{Tr}\{[D_\mu, V][D^\mu, V] - \lambda^2 V^2\}, \tag{2.46}$$

where λ is the mass of the meson.

In the theory containing all the ψ, ϕ, and V fields coupled to the A^a_μ fields, the Lagrangian is equal to the sum of the individual Lagrangians in expressions (2.41), (2.43), and (2.46). Notice that the coupling constants of these fields are given by the same value g. Thus the universality of the coupling constant is a consequence of gauge invariance.

2.4   Quantum Chromodynamics

As we mentioned in chapter 1, we do not expect the proton and the neutron to be elementary. Instead, it is now commonly accepted that the basic constituents of hadrons are quarks [10]. Therefore the fields that appear in the Lagrangian should be the quark fields rather than the fields ψ_p and ψ_n.

Originally, Gell-Mann and Zweig conceived that there were three kinds (or flavors) of quarks: up, down, and strange. (Their fields are denoted by ψ_u, ψ_d, and ψ_s.) The up quark and the down quark form an isospin doublet, whereas the strange quark carries no isospin but has a quantum number called strangeness. Hadrons, such as nucleons, are composed of up quarks and down quarks only, whereas strange particles are composed of strange quarks in addition.

One may ascribe conservation of isospin to the invariance of the Lagrangian under gauge transformations of the first kind between ψ_u and ψ_d (rather than that between ψ_p and ψ_n) and conservation of strangeness to the invariance of the Lagrangian under $U(1)$ gauge transformations of the first kind for ψ_s. One may go even further and postulate that the Lagrangian is invariant under the $SU(3)$ group, which transforms the three-flavor quark fields into linear superpositions of one another. Then, by invoking the Yang-Mills principle, we can obtain gauge mesons that can be regarded as the ones mediating strong interactions.

This picture is unsatisfactory for several reasons. First, the $SU(3)$ symmetry is only an approximate one. For example, the mass of the proton (938.28 MeV) is not equal to that of the Λ baryon (1115.60 MeV), which would be the case if the $SU(3)$ symmetry between the three flavors were exact. Also, conservation of isotopic spin is broken by the electromagnetic interaction, whereas conservation of strangeness is broken by the weak interaction. Second, it is known that there are at least five flavors of quarks, the additional ones being called the charm quark [11] and the bottom quark [12].

It is now known that quarks have, in addition to flavor, another quantum number called the color [13] and that there are three colors, designated

red, white, and blue. Basically, there are three arguments supporting this:
(1) The spin-$\frac{3}{2}$ baryon Ω^- is conceived to be a bound state of three strange
quarks [14]. It is assumed that this bound state can be described by the
Schrödinger equation. It is then found that the quarks have no orbital
angular momentum, and hence the total orbital wave function is completely
symmetric. So is the total spin wave function because the spins of the three
quarks must point in the same direction in order to form a total $\frac{3}{2}$ spin.
This violates the Fermi statistics unless the three strange quarks have three
different colors. (2) The process $e^- e^+ \rightarrow$ hadrons proceeds predominantly
by means of $e^+ e^- \rightarrow \gamma \rightarrow q\bar{q} \rightarrow$ hadrons, where q is a quark and $\bar{q}$ is an
antiquark. Therefore the total cross section for $e^+ e^- \rightarrow$ hadrons is approxi-
mately equal to the total cross section for $e^+ e^- \rightarrow q\bar{q}$ summed over all kinds
of $q\bar{q}$ pairs. The experimental values of $\sigma(e^+ e^- \rightarrow$ hadrons) are roughly a
factor of 3 larger than expected when no color was assumed [15]. (3)
Similarly, the amplitude of $\pi^\circ \rightarrow \gamma\gamma$ is approximated by that of $\pi \rightarrow q + \bar{q} \rightarrow$
$\gamma + \gamma$. The value of the decay rate of $\pi^\circ \rightarrow \gamma\gamma$ is roughly a factor of 9 larger
than expected when no color was assumed [16].

The $SU(3)$ color symmetry is exact. Quantum chromodynamics (QCD)
is the quantum field theory obtained by enforcing the Yang-Mills gauge
symmetry on three-color quark fields. The only difference with the case
discussed in section 2.3 is in the number of fermion fields (three instead of
two). The treatment is almost identical.

Let us denote the three-color quark fields by ψ_1, ψ_2, and ψ_3 and write

$$\psi = \begin{pmatrix} \psi_1 \\ \psi_2 \\ \psi_3 \end{pmatrix}. \tag{2.47}$$

As before, we consider the transformations of ψ represented by

$$\psi \rightarrow \psi' = U\psi, \tag{2.48}$$

where U is a unitary 3×3 matrix. Such matrices form a group known as
$U(3)$. A matrix of this group can be expressed as a phase factor times a
matrix of unit determinant. The phase factors form the group $U(1)$, the
consequence of whose invariance has already been discussed. We therefore
concentrate on the group $SU(3)$, formed by unitary 3×3 matrices with
determinant 1:

$$UU^+ = U^+ U = I, \tag{2.49a}$$

$$\det U = 1. \tag{2.49b}$$

An infinitesimal transformation of this group is given by

$$U = 1 + i\varepsilon\alpha. \tag{2.50}$$

In order to satisfy equations (2.49), α must be Hermitian and traceless. There are eight independent traceless Hermitian 3×3 matrices. We choose them to be [17]

$$\lambda_1 = \begin{pmatrix} 0 & 1 & 0 \\ 1 & 0 & 0 \\ 0 & 0 & 0 \end{pmatrix}, \quad \lambda_2 = \begin{pmatrix} 0 & -i & 0 \\ i & 0 & 0 \\ 0 & 0 & 0 \end{pmatrix}, \quad \lambda_3 = \begin{pmatrix} 1 & 0 & 0 \\ 0 & -1 & 0 \\ 0 & 0 & 0 \end{pmatrix},$$

$$\lambda_4 = \begin{pmatrix} 0 & 0 & 1 \\ 0 & 0 & 0 \\ 1 & 0 & 0 \end{pmatrix}, \quad \lambda_5 = \begin{pmatrix} 0 & 0 & -i \\ 0 & 0 & 0 \\ i & 0 & 0 \end{pmatrix}, \quad \lambda_6 = \begin{pmatrix} 0 & 0 & 0 \\ 0 & 0 & 1 \\ 0 & 1 & 0 \end{pmatrix},$$

$$\lambda_7 = \begin{pmatrix} 0 & 0 & 0 \\ 0 & 0 & -i \\ 0 & i & 0 \end{pmatrix}, \quad \lambda_8 = \begin{pmatrix} \dfrac{1}{\sqrt{3}} & 0 & 0 \\ 0 & \dfrac{1}{\sqrt{3}} & 0 \\ 0 & 0 & -\dfrac{2}{\sqrt{3}} \end{pmatrix}. \tag{2.51}$$

Defining

$$T_a = \tfrac{1}{2}\lambda_a, \qquad a = 1, 2, \ldots, 8, \tag{2.52}$$

we have, similar to equation (2.26),

$$[T_a, T_b] = if_{abc}T_c, \tag{2.53}$$

where the f_{abc} are the structure constants of $SU(3)$ and are totally anti-symmetric. Independent values of f_{abc} are given in table 2.1. We also have

$$\mathrm{Tr}(T_a T_b) = \tfrac{1}{2}\delta_{ab}, \qquad a, b = 1, 2, \ldots, 8. \tag{2.54}$$

Let us again require that the Lagrangian be invariant under independent gauge transformations of $SU(3)$ at every space-time point [3]. As before, we are led to the existence of vector gauge fields. Because there are eight independent T_a, we must introduce eight gauge fields A_μ^a. They are known as color gluon fields. We put, as before,

Table 2.1
Independent structure constants of $SU(3)$

abc	123	147	156	246	257	345	367	458	678
f_{abc}	1	$\dfrac{1}{2}$	$-\dfrac{1}{2}$	$\dfrac{1}{2}$	$\dfrac{1}{2}$	$\dfrac{1}{2}$	$-\dfrac{1}{2}$	$\dfrac{\sqrt{3}}{2}$	$\dfrac{\sqrt{3}}{2}$

$$A_\mu \equiv A_\mu^a T^a, \tag{2.55}$$

$$D_\mu \equiv \partial_\mu + ig A_\mu. \tag{2.56}$$

The equation of transformation for A_μ is again given by equation (2.32), and the Lagrangian

$$\mathscr{L} = \bar{\psi}(i\slashed{D} - m)\psi - \tfrac{1}{4}F_{\mu\nu}^a F^{a\mu\nu} \tag{2.57}$$

is gauge invariant. Equation (2.57) is the Lagrangian for QCD, where

$$F_{\mu\nu}^a = \partial_\mu A_\nu^a - \partial_\nu A_\mu^a - g f_{abc} A_\mu^b A_\nu^c. \tag{2.58}$$

References

[1] N. Bohr, *Naturwissenschaften* 16 (1928): 245, 17 (1929): 483, 18 (1930): 73; N. Bohr and L. Rosenfeld, *Det Kongilege Danske Videnskabernes Selskab* 12 (1933): 8.

[2] W. Heisenberg, *Zeitschrift für Physik* 43 (1927): 172.

[3] C. N. Yang and R. L. Mills, *Physical Review* 96 (1954): 191.

[4] E. Schrödinger, *Annalen der Physik* 79(1926): 361, 489, 80 (1926): 437, 81 (1926): 109.

[5] H. Weyl, *Space, Time, Matter* (New York: Dover, 1950); F. London, *Zeitschrift für Physik*, 42 (1927): 375; H. Weyl, *Zeitschrift für Physik* 56 (1929): 330.

[6] P. A. M. Dirac, *Proceedings of the Royal Society of London* 117 (1928): 610, 118 (1928): 341.

[7] E. Noether, *Nachr. d. König. Gesellsch. d. Wiss. zu Göttingen, Math.-Phys. Klass* (1918): 235–257.

[8] P. A. M. Dirac, *Proceedings of the Royal Society of London* 112 (1926): 661, 114 (1927): 243, 710; P. Jordan and W. Pauli, *Zeitschrift für Physik* 47 (1928): 151; E. Fermi, *Rendiconti della R. Accademia Nazionale del Lincei* 9(1929): 881, 12 (1930): 431; W. Heisenberg and W. Pauli, *Zeitschrift für Physik* 56 (1929): 1, 59 (1930): 168.

[9] W. Heisenberg, *Zeitschrift für Physik* 77 (1932): 1.

[10] M. Gell-Mann, *Physics Letters* 8 (1964): 214; G. Zweig, CERN Reports TH-401, TH-412 (1964).

[11] S. L. Glashow, J. Iliopoulos, and L. Maiani, *Physical Review D* 2 (1970): 1285; J. J. Aubert, U. Becker, P. J. Biggs, J. Burger, M. Chen, G. Everhart, P. Goldhagen, J. Leong, T. McCorriston, T. G. Rhoades, M. Rohde, S. C. C. Ting, S. L. Wu, and Y. Y. Lee, *Physical Review Letters* 33 (1974): 1404; J.-E. Augustin, A. M. Boyarski, M. Breidenbach, F. Bulos, J. T. Dakin, G. J. Feldman, G. E. Fischer, D. Fryberger, G. Hanson, B. Jean-Marie, R. R. Larsen, V. Lüth,

H. L. Lynch, D. Lyon, C. C. Morehouse, J. M. Paterson, M. L. Perl, B. Richter, P. Rapidis, R. F. Schwitters, W. M. Tanenbaum, F. Vannucci, G. S. Abrams, D. Briggs, W. Chinowsky, C. E. Friedberg, G. Goldhaber, R. J. Hollebeek, J. A. Kadyk, B. Lulu, F. Pierre, G. H. Trilling, J. S. Whitaker, J. Wiss, and J. E. Zipse, *Physical Review Letters* 33 (1974):1406; C. Bacci, R. Baldini Celio, M. Bernardini, G. Capon, R. Del Fabbro, M. Grilli, E. Iarocci, L. Jones, M. Locci, C. Mencuccini, G. P. Murtas, G. Penso, G. Salvini, M. Spano, M. Spinetti, B. Stella, V. Valente, B. Bartoli, D. Bisello, B. Esposito, F. Felicetti, P. Monacelli, M. Nigro, L. Paoluzi, I. Peruzzi, G. Piano Mortari, M. Piccolo, F. Ronga, F. Sebastiani, L. Trasatti, F. Vanoli, G. Barbarino, G. Barbiellini, C. Bemporad, R. Biancastelli, M. Calvetti, M. Castellano, F. Cevenini, F. Costantini, P. Lariccia, S. Patricelli, P. Parascandalo, E. Sassi, C. Spencer, L. Tortora, U. Troya, and S. Vitale, *Physical Review Letters* 33 (1974):1408, 1649; M. K. Gaillard, B. W. Lee, and J. L. Rosner, *Reviews of Modern Physics* 47 (1975):277.

[12] S. W. Herb, D. C. Hom, L. M. Lederman, J. C. Sens, H. D. Snyder, J. K. Yoh, J. A. Appel, B. C. Brown, C. N. Brown, W. R. Innes, K. Ueno, T. Yamanouchi, R. J. Fisk, A. S. Ito, H. Jöstlein, D. M. Kaplan, and R. D. Kephart, *Physical Review Letters* 39 (1977):252, 1240.

[13] O. W. Greenberg, *Physical Review Letters* 13 (1964):598; M. Y. Han and Y. Nambu, *Physical Review B* 139 (1965):1006; M. Gell-Mann, *Acta Physica Austriaca*, suppl. 9 (1972):733; H. Fritzsch and M. Gell-Mann, *Proceedings of the Sixteenth International Conference on High Energy Physics* (Batavia, Ill.: National Accelerator Laboratory, 1972), vol. 2, 135.

[14] Y. Eisenberg, *Physical Review* 96 (1954):541; V. E. Barnes, P. L. Connolly, D. J. Crennell, B. B. Culwick, W. C. Delaney, W. B. Fowler, P. E. Hagerty, E. L. Hart, N. Horwitz, P. V. C. Hough, J. E. Jensen, J. K. Kopp, K. W. Lai, J. Leitner, J. L. Lloyd, G. W. London, T. W. Morris, Y. Oren, R. B. Palmer, A. G. Prodell, D. Radojicic, D. C. Rahm, C. R. Richardson, N. P. Samios, J. R. Sanford, R. P. Shutt, J. R. Smith, D. L. Stonehill, R. C. Strand, A. M. Thorndike, M. S. Webster, W. J. Willis, and S. S. Yamamoto, *Physical Review Letters* 12 (1964):204.

[15] A. Litke, G. Hanson, A. Hofmann, J. Koch, L. Law, M. E. Law, J. Leong, R. Little, R. Madaras, H. Newman, J. M. Paterson, R. Pordes, K. Strauch, G. Tarnopolsky, and R. Wilson, *Physical Review Letters* 30 (1973):1189.

[16] R. G. Glasser, N. Seeman, and B. Stiller, *Physical Review* 123 (1961):1014; S. L. Adler, *Physical Review* 177 (1969):2426; J. S. Bell and R. Jackiw, *Nuovo Cimento A*, 60 (1969):47.

[17] M. Gell-Mann, Report CTSL-20, California Institute of Technology (1961); Y. Ne'eman, *Nuclear Physics* 26 (1961):222.

3 Fermion-Fermion Elastic Scattering

3.1 Introduction

In this chapter we study the high-energy behavior of fermion-fermion scattering amplitude in gauge field theories. We make a systematic examination of the Feynman amplitudes for fermion-fermion scattering in both QED and QCD. The amplitudes in QED are somewhat simpler, and we treat them first. In this chapter we also study the especially simple class of diagrams of multi-vector-meson exchange of fermion-fermion scattering in QED. (In these diagrams all vector mesons are emitted by one of the fermions and absorbed by the other.) A discussion of general diagrams for fermion-fermion scattering in QED can be found in chapter 11. The amplitudes in QCD can be obtained from those in QED with but slight modification. The second-order through sixth-order diagrams in QCD are discussed in chapter 12, and the eighth-order and tenth-order diagrams are discussed briefly in chapter 13.

It is well known that the fermion-fermion scattering amplitude in QED has infrared divergence [1]. This divergence is due to the emission of low-energy massless vector mesons. Traditionally, this is handled by introducing an infrared cutoff in the form of a vector meson mass λ [2]. We follow this tradition.

3.2 Notation

Before getting into the details of discussion, it is appropriate to define the notation. The scalar product of two four-vectors p and q is

$$pq = p_\mu q^\mu = p_0 q_0 - \mathbf{p} \cdot \mathbf{q}. \tag{3.1}$$

Also,

$$\not{p} = p_\mu \gamma^\mu = p_0 \gamma_0 - \mathbf{p} \cdot \boldsymbol{\gamma}, \tag{3.2}$$

where the γ_μ are 4×4 Dirac matrices given by

$$\gamma_0 = \gamma^0 = \begin{bmatrix} I & 0 \\ 0 & -I \end{bmatrix}, \qquad \boldsymbol{\gamma} = \begin{bmatrix} 0 & \boldsymbol{\sigma} \\ -\boldsymbol{\sigma} & 0 \end{bmatrix}, \tag{3.3}$$

with I the 2×2 unit matrix and $\boldsymbol{\sigma}$ the 2×2 Pauli matrices

$$\sigma_1 = \begin{bmatrix} 0 & 1 \\ 1 & 0 \end{bmatrix}, \qquad \sigma_2 = \begin{bmatrix} 0 & -i \\ i & 0 \end{bmatrix}, \qquad \sigma_3 = \begin{bmatrix} 1 & 0 \\ 0 & -1 \end{bmatrix}. \tag{3.4}$$

This specific choice of the γ matrices is especially convenient for the study of the high-energy behavior of scattering amplitudes. We use u to denote the four-component spinor that satisfies the Dirac equation for a free fermion of mass m:

$$(E\gamma^0 - \gamma \cdot \mathbf{p} - m)u = 0, \tag{3.5}$$

where $(E, \mathbf{p})$ is the four-momentum of the fermion. Explicitly,

$$u = \sqrt{\frac{E+m}{2m}} \begin{pmatrix} \chi \\ \dfrac{\sigma \cdot \mathbf{p}}{E+m} \chi \end{pmatrix}, \tag{3.6}$$

where χ is an arbitrary two-component spinor. We choose the normalization

$$u^+ \gamma^0 u = 1. \tag{3.7}$$

Then the normalization of χ is

$$\chi^\dagger \chi = 1, \tag{3.8}$$

where the dagger denotes Hermitian adjoint. We also define, as usual,

$$\bar{u} = u^+ \gamma^0.$$

3.3 One-Vector-Meson Exchange

The fermion-fermion elastic scattering amplitude can be studied using perturbation theory. The Feynman rules for the matrix elements $\mathcal{M}$ are shown in figure 3.1. The S-matrix elements are related to $\mathcal{M}$ by

$$S_{fi} = \delta_{fi} + i \frac{(2\pi)^4 \delta^{(4)}(\sum p_f - \sum p_i)}{\prod \sqrt{2\omega_n} \prod \sqrt{E_n/m}} \mathcal{M}_{fi},$$

where ω_n is the energy of an external photon and E_n is the energy of an external fermion. The products are over all external particles.

The lowest-order Feynman diagrams [3] for fermion-fermion scattering [4] are of second order and are shown in figure 3.2, where the four-momenta for the incoming fermions are p_1 and p_2. In the diagrams of figure 3.2, the two incident fermions exchange a vector meson. Thus we call this process that of the one-vector-meson exchange. The scattering amplitude corresponding to the Feynman diagram of figure 3.2a is equal to

Internal photon line	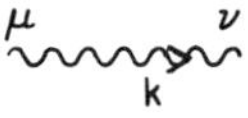 $\mu \qquad \nu$ k	$-\dfrac{ig_{\mu\nu}}{k^2 + i\varepsilon}$
Internal fermion line	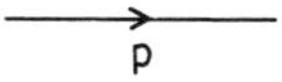 $\longrightarrow$ p	$\dfrac{i}{\not{p} - m + i\varepsilon}$
Vertex	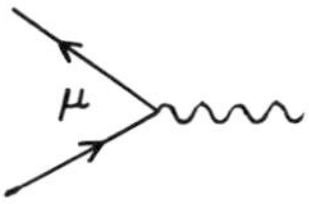 μ	$-ie\gamma_{\mu}$
External incoming fermion line		$u(p)$
External outgoing fermion line		$\bar{u}(p)$
External incoming anti-fermion line		$\bar{v}(p)$
External outgoing anti-fermion line		$v(p)$
External photon line		ε_{μ}
Fermion loop		(-1)
Momentum integration		$\int \dfrac{d^4p}{(2\pi)^4}$
Overall factor		$-i$

Figure 3.1
Feynman rules for QED.

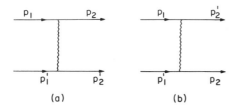

Figure 3.2
The second-order diagrams for fermion-fermion scattering. The solid lines represent fermions, and the wavy lines represent vector mesons.

$$g^2(\bar{u}_2\gamma_\mu u_1)\frac{1}{t - \lambda^2}(\bar{u}_2'\gamma^\mu u_1'),\tag{3.9}$$

where g is the coupling constant and

$$t = (p_2 - p_1)^2\tag{3.10}$$

is the square of the momentum transfer. With the metric (3.1), $t \leqslant 0$ for the physical amplitude, where

$$p_1^2 = p_2^2 = p_1'^2 = p_2'^2 = m^2.$$

For later use, it is useful to define

$$s = (p_1 + p_1')^2,\tag{3.11}$$

the square of the center-of-mass energy. The variables s, t, and $u = 4m^2 - s - t$ are due to Mandelstam.

The amplitude (3.9) for the Feynman diagram of figure 3.2a is exact. For application to hadronic scattering processes, we are interested in its asymptotic form in the high-energy limit of $s \to \infty$, with t fixed at a physical value, that is, a negative value or zero. In this limit the energies of the incident fermions in the center-of-mass system are large, whereas the scattering angle is small. Therefore

$$p_1 \sim p_2 \quad \text{and} \quad p_1' \sim p_2',$$

and a simple calculation with the aid of equation (3.6) gives

$$\bar{u}_2\gamma_\mu u_1 \sim \frac{p_{1\mu}}{m}\chi_2^\dagger\chi_1 \quad \text{and} \quad \bar{u}_2'\gamma_\mu u_1' \sim \frac{p_{1\mu}'}{m}\chi_2'^\dagger\chi_1'.\tag{3.12}$$

By equation (3.8), the factor $\chi_2^\dagger\chi_1$ is equal to 1 when $\chi_2 = \chi_1$, and this means that the helicity of the fermion is conserved. With this understanding of helicity conservation for both fermions in the high-energy limit, we omit the factors $\chi_2^\dagger \chi_1$ and $\chi_2'^\dagger \chi_1'$. With this convention the substitution of relations (3.12) into expression (3.9) shows that the amplitude for one-vector-meson exchange is, asymptotically, in the high-energy limit of $s \to \infty$ with t fixed at a nonpositive value

$$\frac{g^2 s}{2m^2} \frac{1}{t - \lambda^2}. \tag{3.13}$$

Note that in this high-energy limit the contribution from the Feynman diagram of figure 3.2b is smaller by a factor of s and hence is negligible.

We observe that the amplitude (3.13) is real and is equal to s times a function of t. The latter fact means that the contribution of the one-vector-meson exchange amplitude to the total cross section approaches a constant in the high-energy limit.

3.4 Two-Vector-Meson Exchange

We now turn our attention to the next (the fourth-) order amplitude [5]. There are eighteen fourth-order diagrams. Fourteen of these are simply self-energy and vertex insertions to the second-order diagrams discussed in section 3.3. The corresponding amplitudes are simply equal to the second-order amplitude with radiative corrections to the propagators and the vertices. In particular, all of them are equal to s times a function of t and are real. We do not discuss them further. The remaining four diagrams are the two illustrated in figure 3.3 plus the two obtained from them with $p_2 \leftrightarrow p_2'$. As in section 3.3, the latter two diagrams give amplitudes that are too small by a factor of s; they will be neglected. We therefore concentrate on the two diagrams in figure 3.3. We observe that in both of these two diagrams, all virtual vector mesons are emitted by one of the fermions and absorbed by the other one. Thus we call the corresponding process that of two-vector-meson exchange.

Let $\Delta = p_2 - p_1$ be the momentum transfer for the elastic scattering of two fermions so that $t = \Delta^2$. In the center-of-mass system, Δ has no time component, and the space component is in the scattering plane. With reference to figure 3.3, if the momentum carried by one vector meson is designated as q, then that for the other one is $\Delta - q$. We observe that, with

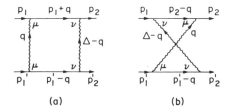

Figure 3.3
The relevant fourth-order diagrams for the absorptive part of fermion-fermion elastic scattering.

the choice shown in figure 3.3, the bottom fermion lines in the two diagrams carry the same momentum. Thus the expressions for the contributions to the scattering amplitude from these two diagrams differ only in the factors associated with the top fermion line. By the Feynman rules, these factors for the diagrams of figures 3.3a and 3.3b are, respectively,

$$\frac{\gamma_\nu(\not{p}_1 + \not{q} + m)\gamma_\mu}{(p_1 + q)^2 - m^2 + i\varepsilon} \tag{3.14a}$$

and

$$\frac{\gamma_\mu(\not{p}_2 - \not{q} + m)\gamma_\nu}{(p_2 - q)^2 - m^2 + i\varepsilon}. \tag{3.14b}$$

Expressions (3.14a) and (3.14b) are exact. Let us next consider these factors in the high-energy limit [6]. Then in the numerators, $\not{q}$, m, and $\not\Delta = \not{p}_2 - \not{p}_1$ are negligible compared with $\not{p}_1$, and in the denominators, q^2 and $q\Delta$ are negligible compared with $p_1 q$. By using relations (3.12), we can apply the following approximations to expression (3.14a):

$$\gamma_\nu(\not{p}_1 + \not{q} + m)\gamma_\mu \sim \gamma_\nu \not{p}_1 \gamma_\mu \sim 2p_{1\mu}\gamma_\nu \sim 2p_{1\nu}p_{1\mu}/m,$$

$$(p_1 + q)^2 - m^2 = 2p_1 \cdot q + q^2 \sim 2p_1 \cdot q \sim 2\omega q_-. \tag{3.15}$$

In relations (3.15) ω denotes the center-of-mass energy of each of the incoming and outgoing fermions. If the z-axis is chosen in the center-of-mass system to be in the direction of $\mathbf{p}_1$, then the q_- of relations (3.15) is defined by

$$q_- = q_0 - q_3. \tag{3.16a}$$

It is also useful to define

$$q_+ = q_0 + q_3. \tag{3.16b}$$

We refer to q_+ and q_- as the longitudinal components of q. As seen later, quantities similar to $q_\pm$ play important roles in the high-energy limit. By relations (3.15), expressions (3.14) are asymptotically

$$\frac{p_{1\nu}p_{1\mu}}{m\omega} \frac{1}{q_- + i\varepsilon}, \tag{3.17a}$$

$$\frac{p_{1\nu}p_{1\mu}}{m\omega} \frac{1}{-q_- + i\varepsilon}, \tag{3.17b}$$

respectively. To obtain the sum of contributions from the two Feynman diagrams of figure 3.3, we add expressions (3.17a) and (3.17b) to get

$$\frac{p_{1\nu}p_{1\mu}}{m\omega} \left(\frac{1}{q_- + i\varepsilon} + \frac{1}{-q_- + i\varepsilon} \right) = -\frac{p_{1\nu}p_{1\mu}}{m\omega} 2\pi i \delta(q_-). \tag{3.18}$$

The Dirac delta function in equation (3.18) means that in the high-energy limit the minus components of the momenta for the two virtual vector mesons can be approximated by 0.

This delta function $\delta(q_-)$ makes it possible to simplify the vector meson propagators in the scattering amplitude

$$q^2 - \lambda^2 \sim -(\mathbf{q}_\perp^2 + \lambda^2),$$

$$(\Delta - q)^2 - \lambda^2 \sim -[(\mathbf{\Delta}_\perp - \mathbf{q}_\perp)^2 + \lambda^2], \tag{3.19}$$

where $\mathbf{q}_\perp$ (the transverse component of q) is

$$\mathbf{q}_\perp = q_1 \mathbf{e}_1 + q_2 \mathbf{e}_2,$$

where $\mathbf{e}_1$ and $\mathbf{e}_2$ are unit vectors in the x direction and the y direction, respectively. The factors associated with the bottom fermion line give approximately

$$\frac{p'_{1\nu}p'_{1\mu}}{m\omega} \frac{1}{-q_+ + i\varepsilon}. \tag{3.20}$$

Because all other factors in the scattering amplitude are independent of q_+, we can symmetrize expression (3.20) and replace $(-q_+ + i\varepsilon)^{-1}$ by

$$\frac{1}{2}\left(\frac{1}{-q_+ + i\varepsilon} + \frac{1}{q_+ + i\varepsilon}\right) = -\pi i \delta(q_+). \tag{3.21}$$

Again, the delta function $\delta(q_+)$ in equation (3.21) means that the plus components carried by the virtual vector mesons are both equal to 0.

We have therefore witnessed an example in which a calculation of Feynman diagrams yields not only the asymptotic amplitudes but also a physical picture with a validity probably exceeding the Feynman diagrams that suggest it. In the scattering process of two-vector-meson exchange, the two fermions retain their large longitudinal momenta and helicities throughout the scattering process, as the vector mesons exchanged are allowed to carry only transverse momenta. The sum of the two-vector-meson exchange amplitudes is easily obtained as

$$is\frac{g^4}{4m^2} \int \frac{d^2 q_\perp}{(2\pi)^2} \frac{1}{(\mathbf{q}_\perp^2 + \lambda^2)[(\mathbf{\Delta}_\perp - \mathbf{q}_\perp)^2 + \lambda^2]}. \tag{3.22}$$

The integral in expression (3.22) is an integral in the momentum space. It is also instructive to transform it into an integral in the configuration space. We therefore introduce the Fourier transform

$$(q_\perp^2 + \lambda^2)^{-1} = \int \frac{d^2 x_\perp}{2\pi} \exp(i\mathbf{q}\cdot\mathbf{x}_\perp)K_0(\lambda|\mathbf{x}_\perp|), \tag{3.23}$$

where K_0 is the modified Bessel function. Using equation (3.23), we transform expression (3.22) into

$$\frac{is}{4m^2} \int d^2 x_\perp \exp(i\mathbf{\Delta}\cdot\mathbf{x}_\perp)\left[\frac{g^2 K_0(\lambda|\mathbf{x}_\perp|)}{2\pi}\right]^2. \tag{3.24}$$

It is interesting to compare the two-vector-meson exchange amplitude with the one-vector-meson exchange amplitude. Expressing the factor $(\Delta_\perp^2 + \lambda^2)^{-1}$ in expression (3.13) by its Fourier transform integral, we can rewrite expression (3.13) as

$$-\frac{s}{2m^2} \int d^2 x_\perp \exp(i\mathbf{\Delta}\cdot\mathbf{x}_\perp)\left[\frac{g^2 K_0(\lambda|\mathbf{x}_\perp|)}{2\pi}\right]. \tag{3.25}$$

Comparing expressions (3.24) and (3.25), we find that, aside from a common factor, the integrand in expression (3.24) is equal to a half of the square of that of expression (3.25).

3.5   Multi-Vector-Meson Exchange

Because the integrand of expression (3.24) for two-vector-meson exchange is proportional to the square of that of expression (3.25) for one-vector-meson exchange, it is natural to ask whether the cube and higher powers of

$$\frac{1}{2\pi} g^2 K_0(\lambda |\mathbf{x}_\perp|)$$

also appear. To answer this question, we study the contribution to the fermion-fermion amplitude from multiple-vector-meson exchange [7,8]. We follow our previous presentation (see [7]).

An n-vector-meson exchange diagram is one with exactly n virtual vector meson lines, each of which is emitted by one of the fermions and absorbed by the other. The total number of n-vector-meson exchange diagrams is equal to $n!$, corresponding to the number of permutations of these vector meson lines. As examples, we illustrate the six diagrams of three-vector-meson exchange in figure 3.4.

Consider an n-vector-meson exchange diagram such as the one illustrated in figure 3.5. Applying the Feynman rules, we get the following factors associated with the top horizontal line:

$$\frac{\gamma_{\mu_n}(\not p_1 + \not q_1 + \not q_2 + \cdots + \not q_{n-1} + m)\gamma_{\mu_{n-1}}\cdots\gamma_{\mu_3}(\not p_1 + \not q_1 + \not q_2 + m)\gamma_{\mu_2}(\not p_1 + \not q_1 + m)\gamma_{\mu_1}}{[(p_1+q_1+q_2+\cdots+q_{n-1})^2 - m^2 + i\varepsilon]\ldots[(p_1+q_1+q_2)^2 - m^2 + i\varepsilon][(p_1+q_1)^2 - m^2 + i\varepsilon]}.$$

$$(3.26)$$

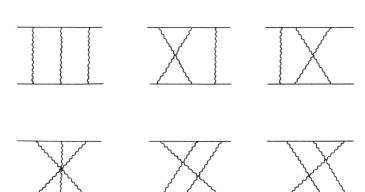

Figure 3.4
The three-vector-meson exchange diagrams for fermion-fermion elastic scattering.

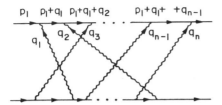

Figure 3.5
A typical diagram of n-vector-meson exchange for fermion-fermion elastic scattering.

As before, we ignore q_i and m as compared to p_1, and $q_i q_j$ as compared to $p_1 q_i$. Expression (3.26) is then reduced to

$$\frac{1}{m} \frac{(p_1)_{\mu_n}(p_1)_{\mu_{n-1}}\cdots(p_1)_{\mu_2}(p_1)_{\mu_1} \delta\left(\sum_{i=1}^{n} q_{i-}\right)}{\omega^{n-1}\left(\sum_{i=1}^{n-1} q_{i-} + i\varepsilon\right)\left(\sum_{i=1}^{n-2} q_{i-} + i\varepsilon\right)\cdots(q_{1-} + q_{2-} + i\varepsilon)(q_{1-} + i\varepsilon)}.$$

$$(3.27)$$

All other n-vector-meson exchange diagrams can be obtained from that in figure 3.5 by permuting the order of vertices on the top fermion line. The scattering amplitudes corresponding to these diagrams differ only in the factors associated with the top fermion line. Thus, referring to expression (3.27), we must calculate, for the sum of all n-vector-meson exchange amplitudes, the sum

$$\frac{\delta\left(\sum_{i=1}^{n} q_{i-}\right)}{\left(\sum_{i=1}^{n-1} q_{i-} + i\varepsilon\right)\left(\sum_{i=1}^{n-2} q_{i-} + i\varepsilon\right)\cdots(q_{1-} + q_{2-} + i\varepsilon)(q_{1-} + i\varepsilon)}$$

+ all other permutations.   (3.28)

For $n = 2$, the sum in expression (3.28) is proportional to the left-hand side of equation (3.18) and has been shown to add up to a delta function, requiring the minus momenta of both vector mesons to vanish. It takes but a little manipulation to show that a more general result applies to arbitrary n and that expression (3.28) is equal to

$$(-2\pi i)^{n-1} \prod_{i=1}^{n} \delta(q_{i-}).$$

$$(3.29)$$

To prove this, we consider the Fourier transform of the first term in expression (3.28):

$$\int \prod_{i=1}^{n} dq_i \exp\left(-i \sum_{i=1}^{n} x_i q_i\right) \frac{\delta\left(\sum_{i=1}^{n} q_i\right)}{\left(\sum_{i=1}^{n-1} q_i + i\varepsilon\right)\left(\sum_{i=1}^{n-2} q_i + i\varepsilon\right)\ldots(q_1 + q_2 + i\varepsilon)(q_1 + i\varepsilon)}$$

$$= \int \prod_{i=1}^{n-1} dq_i \frac{\exp\left(-i \sum_{i=1}^{n-1} (x_i - x_n)q_i\right)}{\left(\sum_{i=1}^{n-1} q_i + i\varepsilon\right)\left(\sum_{i=1}^{n-2} q_i + i\varepsilon\right)\ldots(q_1 + q_2 + i\varepsilon)(q_1 + i\varepsilon)}, \qquad (3.30)$$

where q_{i-} is replaced by q_i for the sake of brevity. If $x_n > x_{n-1}$, we close the contour in the upper half of the complex q_{n-1} plane and find that expressions (3.30) equal 0; and if $x_{n-1} > x_n$, we close in the lower half-plane and reduce expressions (3.30) to

$$(-2\pi i) \int \prod_{i=1}^{n-2} dq_i \frac{\exp\left(-i \sum_{i=1}^{n-2} (x_i - x_{n-1})q_i\right)}{\left(\sum_{i=1}^{n-2} q_i + i\varepsilon\right)\ldots(q_1 + q_2 + i\varepsilon)(q_1 + i\varepsilon)}. \qquad (3.31)$$

Continuing in this manner, we find that expressions (3.30) equal

$$\begin{cases} (-2\pi i)^n & \text{if } x_1 > x_2 > \cdots > x_{n-1} > x_n, \\ 0 & \text{otherwise.} \end{cases} \qquad (3.32)$$

Because expression (3.28) is a sum of all permutations of its first term, we find that, by expressions (3.32), the Fourier transform of expression (3.28) equals $(-2\pi i)^n$ for all x_i, $i = 1, 2, \ldots, n$. Hence expression (3.28) is equal to expression (3.29).

Expression (3.29) simplifies the propagators of the vector mesons:

$$q_i^2 - \lambda^2 \sim -(\mathbf{q}_{i\perp}^2 + \lambda^2), \qquad i = 1, 2, \ldots, n. \qquad (3.33)$$

The factors associated with the bottom fermion line give an expression in the form of expression (3.27), with p_1 replaced by p_1' and the minus momenta replaced by plus momenta. Because all other factors in the scattering amplitude are independent of q_{i+}, $i = 1, \ldots, n$, we can, just as in expression (3.20), symmetricize by adding all its permutations and dividing by $n!$, the number of such terms. Similar to (3.29), we get

$$\frac{(-2\pi i)^{n-1}}{n!} \prod_{i=1}^{n} \delta(q_{i+}). \tag{3.34}$$

Because of expressions (3.29) and (3.34), the sum of n-vector-meson exchange amplitudes in the high-energy limit is equal to an integral over only transverse momenta. By making use of equation (3.23), we can cast this amplitude into the form

$$\mathcal{M}_n \sim -\frac{is}{2m^2} \int d^2 x_\perp \exp(i\mathbf{\Delta} \cdot \mathbf{x}_\perp) \left[-\frac{ig^2 K_0(\lambda |\mathbf{x}_\perp|)}{2\pi} \right]^n \Big/ n!. \tag{3.35}$$

Relation (3.35) is the generalization to arbitrary n of the one-vector-meson exchange amplitude and two-vector-meson exchange amplitude given by expressions (3.25) and (3.24), respectively.

The simplicity of relation (3.35) makes it possible to sum $\mathcal{M}_n$ over n. This total sum of multi-vector-meson exchange amplitudes asymptotically approaches

$$\sum_{n=1}^{\infty} \mathcal{M}_n \sim \frac{is}{2m^2} \int d^2 x_\perp \exp(i\mathbf{\Delta} \cdot \mathbf{x}_\perp) \cdot \left\{ 1 - \exp\left[-\frac{ig^2 K_0(\lambda |\mathbf{x}_\perp|)}{2\pi} \right] \right\}. \tag{3.36}$$

3.6   Discussion

We have gained many interesting insights to high-energy scattering through the calculation of multi-vector-meson scattering amplitudes. We summarize them now.

1. The sum of amplitudes of multi-vector-meson exchange is most elegantly expressed in the form of an integral over $\mathbf{x}_\perp$ as the impact distance; equation (3.36) is reminiscent of the eikonal forms found in high-energy potential scattering (see chapter 4):

$$\int d^2 x \exp(i\mathbf{\Delta} \cdot \mathbf{x}_\perp) \left\{ 1 - \exp\left[-i \int_{-\infty}^{\infty} V(\mathbf{x}_\perp, z)\, dz \right] \right\}, \tag{3.37}$$

where $V(\mathbf{x}_\perp, z)$ is the potential. Comparing equation (3.36) and expression (3.37), we identify $\int_{-\infty}^{\infty} V(\mathbf{x}_\perp, z)\, dz$ with $g^2 K_0(\lambda |\mathbf{x}_\perp|)/2\pi$, the propagator of the vector meson in the configuration space.

2. Fermions retain their helicities and their large longitudinal momenta throughout their scattering process. Indeed, the upper fermion has a plus

momentum approximately equal to 2ω and a minus momentum approximately equal to 0 (or, more precisely, of the order of ω^{-1}) even when it is in a virtual state, and similarly for the lower fermion. The vector mesons exchanged are allowed to carry only transverse momenta. For further discussions, see the references in [9].

3. The results (1) and (2) suggest that it is meaningful to make a distinction between longitudinal momenta and tranverse momenta, which play entirely different roles in high-energy scattering. Mathematically, the high-energy amplitude is best calculated by carrying out all integrations over the longitudinal momenta, with the answer expressed as an integral over the transverse momenta or the impact distances.

4. The eikonal formula (3.36) has been rederived and extended by many authors [10, 11] and has been applied to a wide variety of topics, including the summation of soft pions [12], the Regge model [13], and others [14].

5. The sum of multi-vector-meson amplitudes asymptotically approaches s times a function of Δ. This implies that the elastic differential cross section $d\sigma/dt$ and the total cross section σ both approach constants independent of s as $s \to \infty$. As we will see in chapter 7, this result is dependent on the diagrams we take into consideration and is by no means a general feature.

References

[1] F. Bloch and A. Nordsieck, *Physical Review* 52 (1937): 54; W. Pauli and M. Fierz, *Nuovo Cimento* 15 (1938): 167; S. M. Dancoff, *Physical Review* 55 (1939): 959; H. A. Bethe and J. R. Oppenheimer, *Physical Review* 70 (1946): 451; J. M. Jauch and F. Rohrlich, *Helvetica Physica Acta* 27 (1954): 613; D. R. Yennie, S. C. Frautschi, and H. Suura, *Annals of Physics* (New York), 13 (1961): 379.

[2] R. P. Feynman, *Physical Review* 74 (1948): 1430; J. B. French and V. F. Weisskopf, *Physical Review* 75 (1949): 1240.

[3] R. P. Feynman, *Physical Review* 76 (1949): 769, 80 (1950): 440.

[4] C. Møller, *Annalen der Physik* 14 (1932): 531.

[5] G. Lomanitz, Ph.D. dissertation, Cornell University (1950); M. L. G. Redhead, *Proceedings of the Royal Society of London*, ser. A, 220 (1953): 219.

[6] H. Cheng and T. T. Wu, *Physical Review Letters* 22 (1969): 666, *Physical Review* 182 (1969): 1868, 1899.

[7] H. Cheng and T. T. Wu, *Physical Review* 186 (1969): 1611.

[8] R. Torgerson, *Physical Review* 143 (1966): 1194; M. Levy and J. Sucher, *Physical Review* 186 (1969): 1656; S. J. Chang and S. K. Ma, *Physical Review Letters* 22 (1969): 1334; H. D. I. Abarbanel and C. Itzykson, *Physical Review Letters* 23 (1969): 53; L. N. Chang and N. P. Chang, *Physical Review D* 4 (1971): 1856; G. Feldman and P. T. Matthews, *Journal of Physics A* 6 (1973): 236.

[9] T. C. Meng, *Physical Review D* 6 (1972):1169; W. Czyz and P. K. Kabir, *Physical Review D* 11 (1975):2219; N. S. Khan and V. N. Pervushin, *Theoretical and Mathematical Physics* 29 (1976):1003; M. Quiros, *Helvetica Physica Acta* 50 (1977):81; J. A. McNeil and S. J. Wallace, *Physical Review D* 19 (1979):3145.

[10] D. R. Harrington, *Physical Review D* 4 (1971):840; M. M. Islam, *Nuovo Cimento A* 5 (1971):315; L. N. Chang and G. Segre, *Physical Review D* 6 (1972):2231; L. N. Chang, G. Segre, and N. P. Chang, *Physics Letters B* 39 (1972):207; C. Itzykson, *Annales de Physique* 7 (1972):59; W. Dittrich, *Nuclear Physics B* 45 (1972):290; A. Maheshwari, *Lettere al Nuovo Cimento* 5 (1972):349; A. O. Barut and Z. Z. Aydin, *Nuclear Physics B* 41 (1972):150; M. M. Islam and J. R. Owens, *Physical Review D* 7 (1973):3784; S. J. Wallace, *Physical Review D* 8 (1973):1846; H. Banerjee and S. Mallik, *Physical Review D* 9 (1974):956; W. Dittrich, *Fortschritte der Physik* 22 (1974):539, *Journal of Physics A* 7 (1974):2273; L. Matsson, *Physical Review D* 10 (1974):2010, 2027; H. M. Fried, *Nuclear Physics B* 73 (1974):93; Q. Bui-Duy, *Lettere al Nuovo Cimento* 14 (1975):24; J. P. Harnad, *Annals of Physics* 91 (1975):413; G. P. Lepage, *Physical Review A* 16 (1977):863; S. J. Wallace and J. A. McNeil, *Physical Review D* 16 (1977):3565, 19 (1979):3145.

[11] M. Levy and J. Sucher, *Physical Review D* 2 (1970):1716; R. Blankenbecler and R. L. Sugar, *Physical Review D* 2 (1970):3024; H. M. Fried, *Physical Review D* 3 (1971):2010; E. Eichten, *Physical Review D* 4 (1971):1225; P. Cordero, C. Fronsdal, and T. Gajdicar, *Physical Review D* 5 (1972):2002; W. Dittrich, *Nuclear Physics B* 34 (1971):227, *Physical Review D* 6 (1972):2094, 2104; S. J. Wallace, *Annals of Physics* 78 (1973):190; C. A. Orzalesi, *Proceedings of the 4th International Symposium on Multiparticle Hadrodynamics* (Pavia, Italy, 1973), 564; C. A. Orzalesi, *Lettere al Nuovo Cimento* 10 (1974):85; D. D. Coon and H. Suura, *Physical Review D* 10 (1974):348; M. Doi, *Nuovo Cimento A* 31 (1976):163.

[12] S. Weinberg, *Physical Review D* 2 (1970):674, *Physics Letters B* 37 (1971):494; R. Perrin, *Physical Review D* 3 (1971):1667; I. J. Ketley, T. P. Martin, and J. G. Taylor, *Nuclear Physics B* 34 (1971):567.

[13] S. C. Frautschi, C. J. Hamer, and F. Ravndal, *Physical Review D* 2 (1970):2681; C. J. Hamer and F. Ravndal, *Physical Review D* 2 (1970):2687; R. Torgerson, *Lettere al Nuovo Cimento* 1 (1971):378; J. B. Kogut, *Physical Review D* 4 (1971):3101.

[14] L. F. Li, *Physical Review D* 2 (1970):614; L. Matsson, *Physical Review D* 9 (1974):2894; G. Eilam, *Physical Review D* 13 (1976):942; J. St. Amand and R. A. Uritam, *Physical Review D* 14 (1976):1883.

4 Models for High-Energy Scattering

The principle that can be stated cannot be the absolute principle. The name that can be given cannot be the permanent name.

Lao Tzu

4.1 Role of Potential Models

As shown in chapter 3, the high-energy behavior of many Feynman diagrams is quite easy to obtain. An especially instructive result is that the asymptotic behavior of the scattering amplitude through multi-vector-meson exchange, as given by equation (3.36), is reminiscent of potential scattering.

Before the systematic study of the high-energy behavior in gauge field theory was initiated in the late 1960s, our theoretical knowledge about such behavior came mostly from potential scattering. There were then two distinct schools of thought on the application of potential scattering to the high-energy scattering of strongly interacting particles: the droplet model of Yang and collaborators [1] and the Regge pole model based on crossing [2]. These two models are studied in the next two sections. The intuition of the pioneers working on these two models not only guided our thinking on high-energy processes before the late 1960s but also served as important signposts during the study of gauge field theory itself.

The high-energy behavior of the differential and total cross sections from potential scattering depends on the partial differential equation and the potential used. The crucial difference, however, lies in the kinematics. Consider, for example, the nonrelativistic Schrödinger equation

$$[-\nabla^2 - k^2 + V(\mathbf{x})]\psi(\mathbf{x}) = 0, \tag{4.1}$$

where $\hbar = 2m = 1$ ($\hbar$ is Planck's constant). In the limit $k \to \infty$, we expect the semiclassical WKB approximation [3] to hold. Now, in nonrelativistic classical mechanics, the kinetic energy $\mathbf{p}^2$ of a particle is equal to $k^2 - V(\mathbf{x})$. Furthermore, the classical path traversed by this particle is roughly a straight line if the potential is small compared to k^2. Thus the phase shift resulting from the presence of the potential is approximately the integral of

$$\sqrt{k^2 - V(\mathbf{x})} - k \simeq -\frac{V(\mathbf{x})}{2k}, \qquad k \to \infty, \tag{4.2}$$

over the straight line of the classical motion. Because expression (4.2) vanishes in the limit $k \to \infty$, we expect the scattering cross section to vanish in the limit.

As the second example, consider the Dirac equation

$$(E + i\boldsymbol{\alpha} \cdot \mathbf{V} - m\beta)\psi(\mathbf{x}) = eV(\mathbf{x})\psi(\mathbf{x}), \tag{4.3}$$

where E and m are the energy and the mass, respectively, of the fermion and α and β are the Dirac matrices:

$$\alpha_i = \beta\gamma_i, \qquad i = 1, 2, 3,$$

and

$$\beta = \gamma_0.$$

The incident wave is a plane wave with momentum $\mathbf{P}$:

$$\psi^{\text{inc}}(\mathbf{x}) = e^{i\mathbf{P} \cdot \mathbf{x}}u,$$

where u is a four-component spinor satisfying equation (4.3) with $V = 0$ and

$$|\mathbf{P}| = (E^2 - m^2)^{1/2}.$$

In the semiclassical approximation, the total phase shift along a classical path is the integral of

$$\{[E - eV(\mathbf{x})]^2 - m^2\}^{1/2} - (E^2 - m^2)^{1/2} \simeq -eV(\mathbf{x}), \qquad E \to \infty,$$

independent of E. We therefore expect a constant total cross section at high energies.

This argument gives much more than constant cross section. Without loss of generality, choose the direction of $\mathbf{P}$ to be the z-axis so that $\mathbf{P} = Pe_3$. The classical path of a high-energy particle is then a straight line specified by x and y, or more concisely $\mathbf{x}_\perp$. What we have shown is that the phase shift, denoted by δ, is given by

$$\delta(\mathbf{x}_\perp) = -e \int_{-\infty}^{\infty} dz\, V(\mathbf{x}_\perp, z). \tag{4.4}$$

For the sake of clarity, we have written $\mathbf{x}_\perp$, z instead of $\mathbf{x}$. We emphasize the important role played by relativistic kinematics. Because the scattered wave is proportional to $1 - e^{i\delta}$, the scattering amplitude asymptotically approaches

$$\mathcal{M}(E, \boldsymbol{\Delta}) \sim \frac{iE}{m} \int d^2x_\perp e^{i\boldsymbol{\Delta} \cdot \mathbf{x}_\perp}[1 - e^{i\delta(\mathbf{x}_\perp)}] \tag{4.5}$$

for large E. Here Δ is the momentum transfer; that is, $P\mathbf{e}_3 + \Delta$ is the momentum in the outgoing particle. Because $|P\mathbf{e}_3 + \Delta| = P$ by energy conservation, Δ is perpendicular to $\mathbf{e}_3$ in the limit $E \rightarrow \infty$.

Formula (4.5) shows that *the important variable for high-energy potential scattering is the momentum transfer* Δ [4], not, for example, the scattering angle. Furthermore, the differential cross section of scattering for a given momentum transfer approaches a finite limit as $E \rightarrow \infty$.

These results can be derived in a more mathematical way. Let us define

$$\psi(\mathbf{x}) = e^{iPz}\phi(\mathbf{x}), \tag{4.6}$$

and thus this ϕ satisfies

$$(E - P\alpha_3 + i\boldsymbol{\alpha} \cdot \mathbf{V} - m\beta)\phi(\mathbf{x}) = eV(\mathbf{x})\phi(\mathbf{x}) \tag{4.7}$$

together with the boundary condition $\phi^{\text{inc}}(\mathbf{x}) = u$. Because $\phi(\mathbf{x})$ is not a rapidly varying function of $\mathbf{x}$, $\boldsymbol{\alpha} \cdot \mathbf{V}$ is of order 1. Furthermore, because $E \sim P$ in the limit of high energies, equation (4.7) is simply, to the leading order in E,

$$(1 - \alpha_3)\phi(\mathbf{x}) \sim 0. \tag{4.8}$$

The next step is to multiply the left-hand side of equation (4.7) by $\frac{1}{2}(1 + \alpha_3)$. It follows from the anticommuting properties of $\boldsymbol{\alpha}$ and β that

$$(1 + \alpha_3)O\phi(\mathbf{x}) = O(1 - \alpha_3)\phi(\mathbf{x}) \sim 0 \tag{4.9}$$

for the three cases $O = \alpha_1, \alpha_2$, and β. Therefore, on the left-hand side, only the $\alpha_3(\partial/\partial z)$ term survives, with the result

$$i\frac{\partial}{\partial z}\phi(\mathbf{x}) \sim eV(\mathbf{x})\phi(\mathbf{x}). \tag{4.10}$$

This is a first-order ordinary differential equation whose solution is explicitly

$$\phi(\mathbf{x}_\perp, z) \sim \exp\left[-ie\int_{-\infty}^{z} V(\mathbf{x}_\perp, z')\,dz'\right]u \tag{4.11}$$

because of the boundary condition $\phi^{\text{inc}}(\mathbf{x}) = u$. Relation (4.11) is the desired asymptotic solution for ϕ. Note that the energy E does not appear in relation (4.11).

It follows from relation (4.11) that the scattering amplitude into the final state $e^{i\mathbf{P}' \cdot \mathbf{x}}u'$, where $\mathbf{P}' = \mathbf{P} + \Delta$, is

$$\mathcal{M}(E, \Delta) = -e u'^{\dagger} \int d^3 x \, e^{-i\Delta \cdot x} V(\mathbf{x}) \phi(\mathbf{x})$$

$$\sim -e(u'^{\dagger} u) \int d^3 x \, e^{-i\Delta \cdot x} V(\mathbf{x}_{\perp}, z) \exp\left[-ie \int_{-\infty}^{z} V(\mathbf{x}_{\perp}, z') \, dz' \right].$$

This equation is valid for $E \to \infty$ with Δ kept finite. In this limit, $\Delta_3 \sim 0$, and hence the integration over z can be explicitly carried out to give

$$\mathcal{M}(E, \Delta) \sim i(u'^{\dagger} u) \int d^2 x_{\perp} e^{-i\Delta \cdot x_{\perp}} \left\{ 1 - \exp\left[-ie \int_{-\infty}^{\infty} V(\mathbf{x}_{\perp}, z) \, dz \right] \right\}. \quad (4.12)$$

Relation (4.12) is the same as relation (4.5), as

$$u'^{\dagger}(\mathbf{P} + \Delta) u(\mathbf{P}) \simeq u'^{\dagger}(\mathbf{P}) u(\mathbf{P}) = \bar{u}(\mathbf{P}) \gamma_0 u(\mathbf{P}) = \frac{E}{m} \delta_{fi},$$

where δ_{fi} is the Kronecker delta for the helicity. We can therefore replace δ_{fi} by the unity matrix in the helicity space implied in relation (4.5).

We note that relation (4.5) is a nonperturbative result, in the sense that contributions of the potential to all orders are included. Thus we see that, although it is impossible to solve the Dirac equation explicitly with a general potential at intermediate energies, it requires little effort to do so at high energies. Furthermore, the wave function (4.11) and the scattering amplitude (4.5) are asymptotically in remarkably simple forms. Indeed, when spherical symmetry obtains, the angular integration in relation (4.5) can be carried out to give

$$\mathcal{M}(E, \Delta) \sim \frac{2\pi i E}{m} \int_0^{\infty} x_{\perp} \, dx_{\perp} \, J_0(x_{\perp} \Delta) [1 - e^{i\delta(x_{\perp})}],$$

where $x_{\perp}$ is $|\mathbf{x}_{\perp}|$ and J_0 is the Bessel function of order 0. This is to be compared with the usual partial-wave expansion

$$\mathcal{M}(E, \Delta) = \frac{\pi i}{mE} \sum_{l=0}^{\infty} (2l + 1) P_l(\cos \theta)(1 - e^{2i\delta_l(E)}),$$

where $\theta \ (\sim \Delta/E)$ is the scattering angle, P_l is a Legendre polynomial, and δ_l is the phase shift for the lth partial wave. The point here is that asymptotically for large E, these two formulas are the same, provided that

$$2\delta_l(E) \sim \delta(x_{\perp}) \qquad \text{as } E \to \infty, \qquad\qquad (4.13)$$

where $x_\perp = (l + \frac{1}{2})/E$ is the impact parameter. This follows readily from the asymptotic behavior of the Legendre polynomial

$$P_l(\cos\theta) \to J_0(x_\perp \Delta)$$

as $E \to \infty$ with fixed $x_\perp$ and Δ. Relation (4.13) implies, in particular, that $\delta_l(E)$ is a smooth function of l.

Finally, we list two physical results drawn from relation (4.5).

1. The total cross section σ_{tot} approaches a nonzero limit at infinite energy. The optical theorem [5] gives

$$\sigma_{tot} = \frac{2m}{P} \operatorname{Im} \mathscr{M}(E, 0)\bigg|_{f=i}$$

$$\sim 2 \int d^2 x_\perp \operatorname{Re}\left\{ 1 - \exp\left[-ie \int_{-\infty}^{\infty} V(x_\perp, z)\, dz \right] \right\}$$

independent of E. Moreover, the integrand on the right-hand side in the above equation is nonnegative.

2. The differential cross section $d\sigma/d\Delta$ approaches at infinite energy a limit that is not identically zero. The differential cross section is

$$\frac{d^2\sigma}{d\Delta_1\, d\Delta_2} = \frac{1}{(2\pi)^2} \frac{m^2}{P^2} |\mathscr{M}(E, \Delta)|^2$$

$$\sim (2\pi)^{-2} \delta_{fi} \left| \int d^2 x_\perp e^{-i\Delta \cdot x_\perp} \left\{ 1 - \exp\left[-ie \int_{-\infty}^{\infty} V(x_\perp, z)\, dz \right] \right\} \right|^2,$$

which is a function of Δ but not of E.

4.2   Droplet Model

Although originally based on potential scattering, Yang et al. introduced major modifications in developing their droplet model. As seen explicitly in section 4.1, the high-energy behavior for potential scattering depends on the model used. The physical basis of the droplet model is as follows.

(i) High-energy elastic scattering experiments indicate that, as the incoming energy $E \to \infty$, the differential cross section approaches a limit [1] called $f(t)$:

$$f(t) = \lim_{E \to \infty} \frac{d\sigma}{dt}. \tag{4.14}$$

This in particular implies that both the total cross section and the integrated elastic cross sections approach nonvanishing limits. (The actual situation is somewhat more complicated and is discussed in detail in chapter 7.)

(ii) Above about 300 MeV of excitation energy, the nucleon has many excited states. This excitation energy must be considered to be quite low.

(iii) Since the early 1960s, experiments have shown that the large angle elastic pp cross section drops spectacularly with energy. Over a t range of about 1 $(\text{GeV}/c)^2$, near the forward direction the drop is roughly six orders of magnitude.

Fact (i) indicates that, for a given impact distance $\mathbf{x}_\perp$, the complex phase shift approaches a finite nonzero value in the limit of high energies. Facts (ii) and (iii) suggest that the nucleon is an extended object with an internal structure having a "rigidity" characterized by an excitation energy of the order of a few hundred MeV. For hard collisions in which the available energy is much larger than this, many degrees of freedom are excited in the nucleons, resulting in general in the emission of many particles and hence in a small elastic cross section [6].

Although the picture of the nucleon as an extended object was first conceived for the purpose of understanding hard processes with large momentum transfers, Yang et al. [1] applied this notion also to the case of small momentum transfer. Thus the scattering of two nucleons, or more generally hadrons, is perceived as two objects of finite spatial extension that "go through" each other with attenuation. With the exponent in the eikonal formula taken to be real, application of relation (4.5) gives

$$\mathscr{M}(s,t) \sim \frac{is}{2m^2} \int d^2\mathbf{x}_\perp \exp(i\mathbf{\Delta} \cdot \mathbf{x}_\perp)[1 - e^{-\chi(\mathbf{x}_\perp)}]. \tag{4.15}$$

To determine χ, Yang et al. argued that, to a point inside the incoming particle, the target appears as a disk with a two-dimensional density of opaqueness $D(\mathbf{x}_\perp)$, which is obtained by integrating over the three-dimensional density of opaqueness $\rho(\mathbf{x})$:

$$D(\mathbf{x}_\perp) = \int_{-\infty}^{\infty} \rho(\mathbf{x}_\perp, z)\, dz. \tag{4.16}$$

Therefore, for the collision between particle A and particle B, the resultant opaqueness at an impact distance $\mathbf{x}_\perp$ is

$$\chi(\mathbf{x}_\perp) = K \int D_A(\mathbf{x}_\perp - \mathbf{x}'_\perp) D_B(\mathbf{x}'_\perp) d^2 \mathbf{x}'_\perp, \tag{4.17}$$

where K is a constant.

By the optical theorem, the total cross section is related to $\mathcal{M}$ by

$$\sigma(s) = \frac{4m^2}{s} \, \text{Im} \, \mathcal{M}(s, 0), \tag{4.18}$$

and the differential cross section for elastic scattering is given by

$$\frac{d\sigma}{dt} = \left| \frac{m^2}{s} \mathcal{M}(s, t) \right|^2. \tag{4.19}$$

Substituting equation (4.15) into equations (4.18) and (4.19), we find that, in the limit of infinite energy, $\sigma(s)$ is a constant and $d\sigma/dt$ is a function only of t, both independent of s, in agreement with fact (i). Physically, in the droplet model, a particle at high energies resembles a disk with a size and an opacity both independent of s.

4.3 Regge Pole Model

In 1959, Regge [2] made use of potential scattering as a model for high-energy collisions from a completely different direction. He studied the amplitude of potential scattering in the limit of *fixed energy* and *large* and *complex scattering angle*, that is, in the limit $|\Delta^2| \to \infty$ with s fixed. The outgrowth of such a study is the Regge pole model, which has drastically different physical consequences.

The relevance of Regge's study to high-energy collisions becomes evident as we compare the following two scattering reactions: (1) $a + b \to c + d$, and (2) $a + \bar{c} \to d + \bar{b}$. These reactions are illustrated schematically in figure 4.1. For reaction 1, the center-of-mass energy squared of the system is

$$s = (p_a + p_b)^2, \tag{4.20}$$

and the negative of the momentum transfer squared is

$$t = (p_c - p_a)^2. \tag{4.21}$$

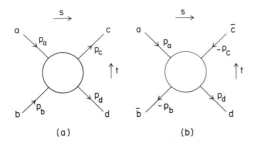

Figure 4.1
(a) Schematic drawing for the reaction $a + b \to c + d$. (b) Schematic drawing for the reaction $a + \bar{c} \to \bar{b} + d$.

Reaction 2 can be obtained from reaction 1 by the replacement

$$p_{\bar{c}} = -p_c \quad \text{and} \quad p_{\bar{b}} = -p_b.$$

Thus for the second reaction the center-of-mass energy squared of the system is

$$(p_a + p_{\bar{c}})^2 = (p_a - p_c)^2 = t, \tag{4.22}$$

and the negative of the momentum transfer squared is

$$(p_{\bar{b}} - p_a)^2 = (p_b + p_a)^2 = s. \tag{4.23}$$

In other words, the Mandelstam variables [4] s and t interchange their roles in these two interactions. It follows that the scattering amplitude of $a + \bar{c} \to d + \bar{b}$, when continued to the region of finite energy and large momentum transfer squared, corresponds to the reaction $a + b \to c + d$ of finite momentum transfer and large energy.

Regge's novel approach for studying the scattering amplitude of large energy with finite momentum transfer is to make use of this correspondence. It is now common to refer to reactions 1 and 2 as those in the s-channel and t-channel, respectively. Thus, instead of studying the high-energy scattering amplitude of finite momentum transfer in the s-channel, Regge studied the low-energy scattering amplitude of large momentum transfer squared in the t-channel. The crucial step that enabled him to do so was made with the recognition that the scattering angle and the angular momentum are conjugate to each other and that, as a consequence, it is useful to study the scattering amplitude as a function of the (complex!) angular momentum.

To explain this, we find it helpful to designate first the kinematics. For simplicity, let us assume that particles a, b, c, and d are spinless and of the same mass m. For the reaction $a + \bar{c} \rightarrow d + \bar{b}$, we denote the center-of-mass energy squared as

$$t = 4(\mathbf{k}^2 + m^2); \tag{4.24}$$

then the negative of the momentum transfer squared is

$$s = -2\mathbf{k}^2(1 - \cos \theta), \tag{4.25}$$

with $\mathbf{k}$ the incident momentum and θ the scattering angle. The limit of $s \rightarrow \infty$ with t fixed is obtained by making θ complex and large in magnitude with $\mathbf{k}^2$ fixed.

It has long been recognized that the asymptotic form of a function as a variable becomes large is related to the singularities of this function in the plane of the conjugate variable. For example, the conjugate of the position variable x is the momentum variable p. The asymptotic form of a function $f(x)$ as $|x| \rightarrow \infty$ is related to the singularities of the Fourier transform of $f(x)$ in the complex plane of p. This relation is easily verified, as the inversion formula for a Fourier transform is well known.

The scattering angle θ is conjugate to the angular momentum l, and consequently we expect the asymptotic form of the scattering amplitude as $|\theta| \rightarrow \infty$ to be related to the singularities of the scattering amplitude in the complex angular momentum plane. Unlike the momentum variable p, however, the physical values of the angular momentum variable l are discrete, and the procedure for obtaining the asymptotic form for large and complex θ is necessarily somewhat different.

We start with the scattering amplitude $f(\mathbf{k}^2, \theta)$ in the partial-wave expansion

$$f(\mathbf{k}^2, \theta) = \sum_{l=0}^{\infty} A_l(\mathbf{k}^2) P_l(\cos \theta), \tag{4.26}$$

where

$$A_l(\mathbf{k}^2) = \frac{(2l + 1)}{k} e^{i\delta_l} \sin \delta_l, \tag{4.27}$$

with δ_l the partial-wave phase shift. If $A_l(\mathbf{k}^2)$, which has physical meaning only when l is equal to a nonnegative integer, can be analytically continued to complex values of l, then we can rewrite equation (4.26) in the form

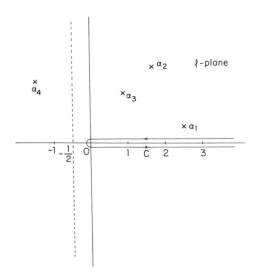

Figure 4.2
A drawing of the complex l-plane, including the contour C for equation (4.28). The singularities of A_l are indicated by the crosses.

$$f(\mathbf{k}^2, \theta) = \frac{1}{2i} \int_C \frac{A_l P_l(-\cos\theta)}{\sin\pi l} \, dl, \tag{4.28}$$

where C is a contour wrapped around the positive real axis of l, enclosing all nonnegative integers and excluding all singularities of A_l, as illustrated in figure 4.2. We can verify equation (4.28) by recognizing that $P_l(-\cos\theta)$ is an entire function of l and that the singularities of the integrand in equation (4.28) enclosed by C are simple poles at the nonnegative integral values. Thus the right-hand side of equation (4.28) is equal to the sum of the residues of the integrand at these values and is precisely the right-hand side of equation (4.26). The minus sign in the argument of the Legendre function in equation (4.26) is due to the relation, valid for integral values of l,

$$P_l(-\cos\theta) = (-1)^l P_l(\cos\theta).$$

Next, we deform the contour C into the line $\operatorname{Re} l = -\frac{1}{2}$. Then equation (4.28) becomes

$$f(\mathbf{k}^2, \theta) = -\sum_{\operatorname{Re}\alpha_n > -\frac{1}{2}} \frac{\pi r_n P_{\alpha_n}(-\cos\theta)}{\sin\pi\alpha_n} - \frac{1}{2i} \int_{-\frac{1}{2}-i\infty}^{-\frac{1}{2}+i\infty} \frac{A_l P_l(-\cos\theta)}{\sin\pi l} \, dl, \tag{4.29}$$

Figure 4.3
Exchange of a particle with spin J and mass μ. The exchange is represented by the wavy line.

where α_n is a pole of A_l in the complex l-plane with r_n the corresponding residue. We call α_n a Regge pole. In equation (4.29) we have assumed that the singularities of A_l are simple poles and that A_l vanishes sufficiently rapidly as $|l| \to \infty$ in the right-hand plane so that the contribution from the infinite contour vanishes. If there are other kinds of singularities, their contributions are of different forms.

Equation (4.29), obtained from equation (4.28) after the Watson-Sommerfeld transformation, leads readily to the asymptotic behavior of the scattering amplitude in the limit $|\cos \theta| \to \infty$ with $\mathbf{k}^2$ fixed. This is because the asymptotic form of $P_\alpha(z)$ as $|z| \to \infty$ is known: It is a linear combination of z^α and $z^{-\alpha-1}$. For $\alpha > -\frac{1}{2}$, we can ignore the term $z^{-\alpha-1}$. Thus equation (4.29) gives

$$f(\mathbf{k}^2, \theta) \sim \sum_{\mathrm{Re}\,\alpha_n > -\frac{1}{2}} \frac{\beta_n(t)s^{\alpha_n(t)}}{\sin \pi \alpha_n(t)}, \qquad |s| \to \infty, \, t \text{ fixed}, \tag{4.30}$$

where β_n is independent of s. The integral in equation (4.29), called the background term, vanishes at least as fast as $s^{-1/2}$.

Relation (4.30) gives the explicit dependence of the high-energy amplitude in the s-channel on the Regge poles in the t-channel. We see that this amplitude is given by a sum of powers of s, with the exponent of these powers equal to the location of the Regge poles α_n. If A_l has singularities other than simple poles, logarithmic factors of s may also appear.

It is instructive to compare relation (4.30) with the amplitude for the exchange of a particle as illustrated in figure 4.3. Let this particle be of spin J and mass μ; then as a consequence of the conservation of angular momentum in the t-channel, this contribution is necessarily of the form

$$g^2 \frac{P_J(\cos \theta)}{t - \mu^2},$$
(4.31)

where g is the coupling constant and $(t - \mu^2)^{-1}$ is the propagator for the particle exchanged. In the limit $s \to \infty$ with t fixed, expression (4.31) is proportional to

$$g^2 \frac{s^J}{t - \mu^2}.$$
(4.32)

Comparing expressions (4.30) and (4.32), we see that relation (4.30) is a generalization of expression (4.32). More precisely, expression (4.32) corresponds to the contribution of a Regge pole fixed at $\alpha(t) = J$ for all t.

To account for the fact that the total cross sections for hadron-hadron scattering are approximately constants in the high-energy limit, Chew and Frautschi [7] postulated that the Regge pole that dominates elastic scattering satisfies

$$\alpha(0) = 1.$$
(4.33)

From the Schrödinger equation, it was learned that $\alpha(t)$ increases with increasing t. Thus, for small t,

$$\alpha(t) \approx 1 + at.$$
(4.34)

With relations (4.33) and (4.34),

$$\frac{d\sigma}{dt} \propto s^{-2a|t|} = e^{-2a|t| \ln s}.$$
(4.35)

Physically, equation (4.35) says that a hadron resembles a disk with a radius increasing as $\sqrt{\ln s}$. To keep the total cross section a constant, this disk becomes increasingly transparent as the radius increases with the energy.

There are several review articles on these and other models for high-energy scattering [8].

References

[1] N. Byers and C. N. Yang, *Physical Review* 142 (1966):976; T. T. Chou and C. N. Yang, in *High Energy Physics and Nuclear Structure* (Amsterdam: North-Holland, 1967), 348–359; T. T. Chou and C. N. Yang, *Physical Review Letters* 20 (1968):1213, *Physical Review* 170 (1968):1591, 175 (1968):1832; J. Benecke, T. T. Chou, C. N. Yang, and E. Yen, *Physical Review* 188 (1969):2159; T. T. Chou and C. N. Yang, *Physical Review Letters* 25 (1970):1072; J. Benecke, *Nuovo Cimento A* 2 (1971):615; J. P. Hsu, *Physical Review D* 3 (1971):257.

[2] T. Regge, *Nuovo Cimento* 14 (1959):951, 18 (1960):947. There are a number of books on Regge poles: S. Frautschi, *Regge Poles and S-Matrix Theory* (London: Benjamin, 1963); R. Omnes and M. Froissart, *Mandelstam Theory and Regge Poles* (London: Benjamin, 1963); E. J. Squires, *Complex Angular Momenta and Particle Physics* (London: Benjamin, 1963).

[3] G. Wentzel, *Zeitschrift für Physik* 38 (1926):518; H. A. Kramers, *Zeitschrift für Physik* 39 (1926):828; L. Brillouin, *Journal de Physique* 7 (1926):353.

[4] S. Mandelstam, *Physical Review* 112 (1958):1344.

[5] N. Bohr, R. Peierls, and G. Placzek, *Nature* 144 (1939):200.

[6] T. T. Wu and C. N. Yang, *Physical Review* 137 (1965):B708.

[7] G. F. Chew and S. Frautschi, *Physical Review Letters* 5 (1960):580, 7 (1961):394, *Physical Review* 123 (1961):1478; R. Blankenbecler and M. L. Goldberger, *Physical Review* 126 (1962):766; S. C. Frautschi, M. Gell-Mann, and F. Zachariasen, *Physical Review* 126 (1962):2204.

[8] F. Zachariasen, *Physics Reports C* 2 (1971):1; W. R. Frazer, L. Ingber, C. H. Mehta, C. H. Poon, D. Silverman, K. Stowe, P. D. Ting, and H. J. Yesian, *Reviews of Modern Physics* 44 (1972):284; R. J. N. Phillips, *Proceedings of the Royal Society of London*, ser. A, 335 (1973):485.

5 Multiparticle Production

Liu Bang (the emperor who overthrew the Chin dynasty) often chatted with Han Hsin, his brilliant field marshal, about his generals. Liu liked to rate the generals by saying how many soldiers each of them was capable of commanding. One day the emperor finally asked Han Hsin, "How many men do you think I am capable of commanding?" Han Hsin answered, "Your majesty can command no more than ten thousand." "Then, how many can you command?" Han Hsin replied, "For your humble servant, the more the better."

History of Han

5.1 Introduction

In a scattering process the energy of the system is a conserved quantity. As a consequence, particle production cannot take place if the kinetic energy available is less than the rest mass of any particle. Therefore, in the collision of two low-energy particles, the only process that can happen is elastic. For such low-energy processes we may expect potential scattering to be a reasonable description.

The situation is entirely different for high-energy scattering. Let us consider the scattering of two particles in the center-of-mass frame. For the sake of simplicity, let the masses of these two particles be equal, and let the energy of each incident particle be ω. Thus s, the square of the center-of-mass energy, is equal to $4\omega^2$. As ω increases, it becomes kinematically possible to create a higher and higher number of particles. Furthermore, the energy of these created particles has a larger and larger range. Therefore we expect a great number of particle production processes to take place in high-energy scattering.

The scattering matrix is unitary—a consequence of the fact that the total probability that the system will end up in some state is 1. Therefore we have, for example, the optical theorem relating the imaginary part of the elastic scattering amplitude in the forward direction to the total cross section. As a result, the existence of the production processes influences even the behavior of the elastic scattering amplitude. To study such interconnection, we cannot stay with potential scattering, as this is inadequate. We must look to relativistic quantum field theories for answers.

At first sight, a description of a high-energy scattering process with many particles produced seems hopelessly complicated. Actually, there is a great deal of regularity. As we have mentioned before, the phase space available for a produced particle is large. Because this available phase space is clearly not uniformly occupied, the question can naturally be raised, What final momenta are favored?

An important clue to the answer can be found from the results of chapter 3. For fermion-fermion scattering at high energies, the amplitude decreases rapidly as the momentum transfer, or, equivalently, the transverse momentum of each of the outgoing particles increases. The reason is simply that, for a larger momentum transfer to take place, at least one of the exchanged vector mesons must go far away from its mass shell (that is, if q is the momentum of the vector meson, $q^2 - \lambda^2$ becomes large). Because this factor appears in the denominator of the scattering amplitude, the scattering amplitude becomes small.

This condition clearly holds also for production processes. More precisely, at high energies the important region of the phase space has the property that the transverse momentum of each of the outgoing particles is not large. We therefore focus our attention on this phase space region for the time being and refer, for convenience, to the *limit of $s \to \infty$ with all transverse momenta fixed as the high-energy limit.*

It is more interesting to consider the dependence of the scattering amplitude on the longitudinal momentum. By energy conservation, k_z of each of the outgoing particles is bounded approximately by

$$-\omega < k_z < \omega. \tag{5.1}$$

It will prove convenient to define

$$x = \frac{k_z}{\omega}. \tag{5.2}$$

Then the range (5.1) includes

a. x is a positive fraction of order 1

b. x is a negative fraction of order 1 and

c. x is small.

The distinction of case c from cases a and b is of great importance. We use the term *pionization* to refer to a process in which a particle with small x is produced, and such produced particles are called *pionization products* [1]. In the next chapter the influence of pionization on elastic scattering through unitarity is studied in detail. In contrast, outgoing particles with positive (resp. negative) fractional x of order 1 are called fragments of the incoming particle moving in the $+z$ (resp. $-z$) direction [2].

The physical meaning of fragments and pionization products becomes clearest if we define

$$k_{\pm} = k_0 \pm k_3;$$

thus

$$k^2 = k_+ k_- - \mathbf{k}_{\perp}^2 \tag{5.3a}$$

$$2p \cdot k = p_- k_+ + p_+ k_- - 2\mathbf{p}_{\perp} \cdot \mathbf{k}_{\perp}. \tag{5.3b}$$

For $x > 0$, we have from equation (5.2) that

$$k_+ \simeq 2\omega x, \qquad x > 0. \tag{5.4a}$$

Because $k^2 = \lambda^2$, we have from relations (5.3a) and (5.4a) that

$$k_- = \frac{\mathbf{k}_{\perp}^2 + \lambda^2}{k_+} \simeq \frac{\mathbf{k}_{\perp}^2 + \lambda^2}{2\omega x}, \qquad x > 0. \tag{5.4b}$$

(The sum $\mathbf{k}_{\perp}^2 + \lambda^2$ is called the square of the transverse mass.) In other words, k_+ is large and k_- is small for $x > 0$. Similarly, for $x < 0$, we have

$$k_- \simeq -2\omega x, \qquad x < 0, \tag{5.4c}$$

$$k_+ \simeq -\frac{\mathbf{k}_{\perp}^2 + \lambda^2}{2\omega x}, \qquad x < 0. \tag{5.4d}$$

In other words, k_- is large and k_+ is small for $x < 0$. The plus and the minus momenta of the incident particle moving in the positive z direction can be similarly obtained as

$$p_{1+} \simeq 2\omega, \qquad p_{1-} \simeq \frac{m^2}{2\omega}.$$

The plus, the minus, and the transverse components transform simply under Lorentz transformations in the z direction. Under these transformations, $\mathbf{k}_{\perp}$ is invariant, whereas

$$k'_+ = \sqrt{\frac{1-\beta}{1+\beta}}k_+, \qquad k'_- = \sqrt{\frac{1+\beta}{1-\beta}}k_-, \tag{5.5}$$

where β is the velocity parameter of the Lorentz transformation.

Let us view the particle with momentum k in the rest frame of one of the incoming fermions, say the one moving in the $+z$ direction. The Lorentz transformations relating the components k'_+, k'_-, $\mathbf{k}'_\perp$ of k in this rest frame to its components k_+, k_-, $\mathbf{k}_\perp$ in the center-of-mass system are

$$k'_+ = k_+ \frac{m}{2\omega}, \qquad k'_- = k_- \frac{2\omega}{m}, \qquad \mathbf{k}'_\perp = \mathbf{k}_\perp. \tag{5.6}$$

Substituting relations (5.4a) and (5.4b) into equations (5.6), we get

$$k'_+ \simeq xm, \qquad k'_- \simeq \frac{\mathbf{k}_\perp^2 + \lambda^2}{xm}, \qquad x > 0.$$

Thus, if x is a finite positive fraction, both k'_+ and k'_- are of the order of 1. This says that a fragment of a fermion is of finite energy in the rest frame of this fermion. In the limit $x \to 0^+$, k'_- becomes infinite, whereas k'_+ vanishes. This says that a pionization product has a large and negative k'_z in the rest frame of the incident fermion in question. Needless to say, the fragment of the other incident fermion (moving in the $-z$ direction) has an even larger k'_-. Indeed, by substituting relation (5.4c) into the middle equation (5.6) for k'_-, we get

$$k'_- \simeq |x|s/m, \qquad x < 0,$$

in the rest frame of the fermion in question.

The consideration above is purely kinematic. It does not tell us, for example, whether fragments or pionization products are favored. Indeed, from this consideration it is not even clear whether the cross section for creating any of them is nonvanishing in the limit $\omega \to \infty$. To answer such important questions of dynamics, we make specific studies of inelastic processes in quantum electrodynamics.

Let f denote a fermion, $\bar{f}$ its antiparticle, and V a vector meson. A simple production process is

$$f + f \to f + f + V, \tag{5.7a}$$

and a slightly more complicated one is

$$f + f \to f + f + f + \bar{f}. \tag{5.7b}$$

Throughout this book, the vector meson V is massive unless otherwise explicitly stated. The process (5.7a) is conveniently referred to as *bremsstrahlung* and is studied in section 5.2 to the lowest order. The pair-production process (5.7b) is studied in section 5.3 to its lowest order.

Before we get into the details of diagrammatic studies, let us reiterate the following simple fact: In a Feynman diagram an internal line of momentum q and mass μ contributes a factor $(q^2 - \mu^2)^{-1}$ to the scattering amplitude. If q^2 is large (in the scale of the rest masses of the particles), we say that this line is far off the mass shell and that the propagator is small. If q^2 is of the order of μ^2, we say that this line is near the mass shell and that the propagator is of the order of 1. If $q^2 = \mu^2$, we say that this line is on the mass shell and that the propagator is equal to infinity.

We require that no internal lines in a diagram be far off the mass shell. Otherwise the corresponding scattering amplitude is small and negligible.

The magnitude of q^2 is best seen by expressing it as a function of q_+ and q_- as in equation (5.3a). It follows that, if q_+ *is large,* q_- is required to be small, and vice versa. Therefore no internal line may carry *both* a large plus momentum *and* a large minus momentum. More precisely, if the plus momentum of a line is of the order of ω, then the minus momentum of this line must be of the order of ω^{-1}.

5.2   Bremsstrahlung

For the *bremsstrahlung* process (5.7a), there are, to the lowest order, four diagrams, as shown in figure 5.1 [3]. Actually, four more diagrams of the

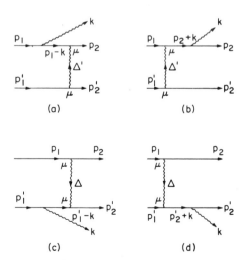

Figure 5.1
Lowest-order diagrams for the *bremsstrahlung* process for fermion-fermion scattering.

same order can be drawn by interchanging the two outgoing fermions, but these additional diagrams are of no importance in the limit of $s \to \infty$ with fixed transverse momenta, in much the same way that the diagram of figure 3.2b for elastic scattering is unimportant. In both the case of these four additional diagrams and the case of figure 3.2b, the exchanged vector meson is far off the mass shell.

Consider the diagram in figure 5.1a, particularly the line with momentum $(p_1 - k)$. We require that $(p_1 - k)^2$ be of the order of 1. From relations (5.4c) and (5.4d), we find that this requirement is not met if $x < 0$, as both the plus and the minus components of $(p_1 - k)$ are large. On the other hand, from relations (5.4a) and (5.4b), we find that this requirement is met if $x > 0$, as the minus component of $(p_1 - k)$ is as small as ω^{-1}. This simple kinematic fact has far-reaching consequences. If the produced vector meson V carries a positive fraction of the momentum of the incoming fermion that emits V itself, then the amplitude is important in the high-energy limit; on the other hand, if V carries a positive fraction of the momentum of the other incoming fermion, then the amplitude is negligible in the high-energy limit. In other words, in the high-energy limit for *bremsstrahlung* processes, the emitted particle (vector meson) and the emitting particle (fermion) must have momenta of the same nature (that is, either both with large plus components and small minus components or both with large minus components and small plus components). Thus, in the lowest-order *bremsstrahlung* process, the emitted meson is always a fragment.

It remains to study the limit of $x \to 0^+$. The propagator for the line with momentum $(p_1 - k)$ is proportional to

$$[(p_1 - k)^2 - m^2]^{-1} = [-2p_1 k + \lambda^2]^{-1}. \tag{5.8}$$

Because from equation (5.3b)

$$2p_1 k = p_{1+} k_- + p_{1-} k_+,$$

we get, making use of relations (5.4a) and (5.4b),

$$2p_1 k \simeq (\mathbf{k}_\perp^2 + \lambda^2)/x + xm^2, \qquad x > 0.$$

The right-hand side is free of ω; thus x is a good variable to use. We call x the fractional longitudinal momentum. As $x \to 0^+$, the quantity $2p_1 k$ blows up as x^{-1}. Thus the propagator vanishes as x. This shows that the processes in figure 5.1 yield no particles with small fractional longitudinal momenta.

People who are aware of the infrared behavior of the *bremsstrahlung* process may wonder how to reconcile this result with the well-known fact that a great number of soft (low-energy) photons are emitted in electron-electron scattering. The crucial point is that a photon is massless, whereas the V particle is massive. Consider again the propagator given by equation (5.8). As $|\mathbf{k}| \to 0$, it is approximately equal to $(-2\omega\lambda + \lambda^2)^{-1}$, which is far off the mass shell, vanishing as $\omega \to \infty$. Thus the amplitude for emitting a V particle with zero momentum is small. However, if $\lambda = 0$, this propagator is approximately $-2\omega|\mathbf{k}|(1 - \cos\theta)$ if $\theta \neq 0$. This is on the mass shell and blows up as $\mathbf{k} \to 0$. Therefore the amplitude for emitting a soft photon is large. (For $\lambda = 0$ and for $\mathbf{k}$ of order 1, we find that, at $\theta = 0$, $(-2p_1 k)^{-1}$ is approximately equal to $-\omega/(m^2|\mathbf{k}|)$, which is larger than its value at $\theta \neq 0$ by a factor of ω^2/m^2. There is therefore a strong forward peak with a magnitude proportional to ω^2! Because the solid angle of the peak is of the order of m^2/ω^2, the energy distribution of the photon, obtained by integrating over the direction of $\mathbf{k}$, approaches a nonzero limit as $\omega \to \infty$ and is of the form dk/k.)

The rest of this section is devoted to a more detailed study of *bremsstrahlung*. As we have discussed, the *bremsstrahlung* amplitude corresponding to the sum of the two diagrams of figures 5.1a and 5.1b is negligible if $x < 0$. Therefore we study this amplitude only for $x > 0$, where this amplitude asymptotically approaches

$$\mathcal{M} \simeq -\frac{g^3}{(\Delta_\perp'^2 + \lambda^2)m}\left[\frac{p_1'(p_1 - k + m)\ell}{(p_1 - k)^2 - m^2} + \frac{\ell(p_2 + k + m)p_1'}{(p_2 + k)^2 - m^2}\right]. \tag{5.9}$$

In relation (5.9), ε is the polarization vector of the produced meson satisfying

$$\varepsilon k = 0,$$

and we have made use of relations (3.12). The numerators in relation (5.9) can be calculated by moving p_1 to the right and p_2 to the left. For example:

$$p_1'(p_1 - k + m)\ell = p_1'(2p_1\varepsilon + \ell k) \simeq p_1'(2p_1\varepsilon + xm\ell + \ell k_\perp),$$

where $k_\perp \equiv -\mathbf{k}_\perp \cdot \gamma_\perp$. Let ε be transverse; then the transverse components of ε are of the order of 1, whereas the longitudinal components of ε are of the order of ω^{-1}. (If ε is longitudinal, the longitudinal components of ε are of the order of $x\omega$. However, we can invoke gauge invariance and replace ε by $\varepsilon - (k/\lambda)$, which has small longitudinal components.) Thus $2p_1\varepsilon + xm\ell + \ell k_\perp$ is of the order of 1. Also, p_1' is of the order of s (see relations (3.12)). Thus the numerators in relation (5.9) are of the order of s. As we have mentioned, the denominators in relation (5.9) are of the order of 1. Thus relation (5.9) is of the order of s. We can then express the scattering amplitude in the form

$$\mathcal{M} \simeq \frac{s}{m^2} F(x, \mathbf{k}_\perp).$$

By dividing $\mathcal{M}$ by the kinematic factor (a factor ω/m for each external fermion and a factor $2E$ for each external V particle, where E is the energy of the V particle) and taking the absolute value squared, we obtain after some algebra the differential cross section for *bremsstrahlung*:

$$d\sigma \simeq \frac{1}{\pi} |F(x, \mathbf{k}_\perp)|^2 \frac{dx}{x} \frac{d^2 k_\perp}{(2\pi)^2} \frac{d^2 p_{2\perp}}{(2\pi)^2}, \qquad x > 0, \tag{5.10}$$

where we have approximated dk_3/E by dx/x. Therefore the lowest-order differential cross section for emitting a fragment V approaches a limit independent of ω as $\omega \to \infty$. As $x \to 0$, F vanishes as x. Therefore the differential cross section (5.10) is proportional to

$$x^2 \frac{dx}{x} = x\,dx, \qquad x \ll 1,$$

which vanishes at $x = 0$.

5.3   Pair Production

We have concluded that the lowest-order *bremsstrahlung* processes yield fragments but not pionization products. We ask, What processes, if any, can yield pionization products?

To answer this question, we observe that the momentum of a pionization product in the center-of-mass system is much smaller than $s^{1/2}$. Thus both k_+ and k_- are much larger than $s^{-1/2}$. From the discussion in section 5.2, a pionization product, or any collection of pionization products, cannot be emitted by any one of the fermions alone. The simplest process for the production of a pionization product is therefore the following: Each of the fermions emits a vector meson, one of which carries a plus momentum but not a minus momentum, and the other carries a minus momentum but not a plus momentum. These two vector mesons annihilate and turn

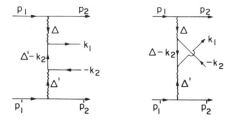

Figure 5.2
The lowest- (fourth-) order diagrams in which the created fermion-antifermion pair may
appear as pionization products.

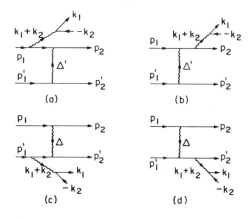

Figure 5.3
Other fourth-order diagrams of fermion-fermion scattering in which a fermion-antifermion
pair is created.

into one or more particles that, by momentum conservation, possess both
a plus momentum and a minus momentum and are hence pionization
products [1, 4]. The lowest-order diagrams which contain this process are
illustrated in figure 5.2. In these diagrams the pionization product is not a
single particle but a fermion-antifermion pair. This slight complication has
no deep dynamic implication and arises merely because two vector mesons
cannot turn into a single vector meson or a single fermion.

The reaction described in figure 5.2 is the pair-production process (5.7b).
For this reaction there are four more diagrams of the same (lowest possible)
order, and we illustrate them in figure 5.3. We recognize that in all the
diagrams in figure 5.3 the pair is attached to one vector meson only. Thus

we expect that the pair in these four diagrams can form fragments but not pionization products. Therefore the lowest-order diagrams yielding pionization products are those in figure 5.2.

Explicit calculations confirm these expectations. In particular, let us consider the diagrams in figure 5.2 in the kinematic region where the two produced particles are both pionization products with energies of the order of 1. Referring to figure 5.2, we have

$$\Delta + \Delta' = k_1 + k_2. \tag{5.11}$$

Because both p_{1-} and p_{2-} are small, we have

$$\Delta_- \sim 0. \tag{5.12}$$

Similarly,

$$\Delta'_+ \sim 0. \tag{5.13}$$

Relations (5.11), (5.12), and (5.13) imply that

$$\Delta_+ \sim k_{1+} + k_{2+}, \tag{5.14}$$

$$\Delta'_- \sim k_{1-} + k_{2-}. \tag{5.15}$$

The scattering amplitude corresponding to the diagrams in figure 5.2 can be calculated as follows. Each of the two vertices on the incident fermions gives a factor ω/m (see relations (3.12)). Also, because of relations (5.12) and (5.13), we have

$$(\Delta^2 - \lambda^2)^{-1} \sim -(\Delta_\perp^2 + \lambda^2)^{-1}, \tag{5.16}$$

$$(\Delta'^2 - \lambda^2)^{-1} \sim -(\Delta_\perp'^2 + \lambda^2)^{-1} \tag{5.17}$$

independent of s. The factors associated with the created pair give the expression

$$F = \frac{\gamma_-(\rlap{/}{\Delta}'_\perp - \rlap{/}{k}_{2\perp} + m)\gamma_+}{k_{1-}k_{2+} + (\Delta'_\perp - \mathbf{k}_{2\perp})^2 + m^2} + \frac{\gamma_+(\rlap{/}{\Delta}_\perp - \rlap{/}{k}_{2\perp} + m)\gamma_-}{k_{2-}k_{1+} + (\Delta_\perp - \mathbf{k}_{2\perp})^2 + m^2}, \tag{5.18}$$

where the first and the second terms come from figures 5.2a and 5.2b, respectively. In equation (5.18), we have simplified the numerators by using the identity

$$\gamma_+^2 = \gamma_-^2 = 0$$

and have adopted the notation

$$\not{p}_\perp \equiv -\gamma_\perp \cdot \mathbf{p}_\perp. \tag{5.19}$$

Equation (5.18) is independent of s. Thus the pair-production amplitude in the pionization region is equal to

$$\frac{g^4 \omega^2 m^{-2} F(k_1, k_2, \Delta)}{(\Delta_\perp^2 + \lambda^2)(\Delta_\perp'^2 + \lambda^2)}, \tag{5.20}$$

which is a function of s times a function of Δ, k_1, and k_2.

We mention two features for the fermion pair production: (1) In the high-energy limit, the pair must have comparable plus (or, equivalently, minus) momenta. For example, in the limit $k_{1+}/k_{2+} \to 0$, a little algebra shows that the first term in equation (5.18) vanishes because its denominator becomes infinite, and the second term in equation (5.18) vanishes because its numerator goes to 0. This is also true for the limit $k_{2+}/k_{1+} \to 0$. (2) Because pionization is due to the interaction of two vector mesons, emitted respectively by the two incident particles, it is natural to introduce the concept of the equivalent vector meson beams, which are used to replace the incident particle beams. This is known as the Weizsäcker-Williams approximation [5]. More precisely, the Weizsäcker-Williams approximation is applied twice, once for each vector meson. Compared with the single Weizsäcker-Williams approximation, such double Weizsäcker-Williams approximations have subtleties in connection with polarization [5].

Additional discussions of particle production can be found in various articles [6, 7].

5.4 Discussion of Pionization

Because the pair-production amplitude of figure 5.2 is linear in s, the lowest-order differential cross section for pionization, equal to the absolute value squared of this amplitude divided by a kinematic factor (which is proportional to s^2 in the high-energy limit), is independent of s. If we integrate this differential cross section over all momenta except one, say $\mathbf{k}_1$, we obtain the one-particle distribution for pionization products in the form

$$\frac{d\sigma}{d\mathbf{k}} \simeq G(k), \tag{5.21a}$$

valid in the region

$$k_+ \ll \omega, \qquad k_- \ll \omega, \tag{5.21b}$$

where we have omitted the subscript 1 for simplicity.

We emphasize that G in relation (5.21a) is independent of s and is a function of $\mathbf{k}$ only. It is further possible to deduce the dependence of $G(k)$ on k_z [1]. To do this, we observe that the scattering amplitude we use to derive equation (5.21a) is invariant under Lorentz transformations in the z direction (although it is no longer invariant under Lorentz transformations in the x or y direction because we have already specified the z-axis to be in the direction of the incident momenta). Thus the differential cross section $d\sigma$ must be invariant under Lorentz transformations in the z direction. It follows from equations (5.5) that, under these transformations, dk_+/k_+ and dk_-/k_- are both invariant under Lorentz transformations in the z direction. A little algebra gives

$$\frac{dk_+}{k_+} = -\frac{dk_-}{k_-} = \frac{dk_z}{E} = \frac{1}{2} d \ln \left(\frac{k_+}{k_-} \right). \tag{5.22}$$

Thus the most general form for $d\sigma$ is

$$d\sigma = f(\mathbf{k}_\perp) \frac{d^3 k}{E}, \tag{5.23}$$

valid in the region (5.21b). In equation (5.23), $E = \sqrt{\mathbf{k}^2 + m^2}$ is the energy of the particle in question. Because the function f in equation (5.23) depends on $\mathbf{k}_\perp$ only, the longitudinal momentum dependence of $d\sigma$ is completely determined.

The function

$$\eta \equiv \frac{1}{2} \ln \left(\frac{k_+}{k_-} \right) \tag{5.24}$$

in equation (5.22) is called the rapidity. Equation (5.23) can also be written in terms of η:

$$d\sigma = f(\mathbf{k}_\perp) d^2 k_\perp \, d\eta, \qquad k_+ \ll \omega, \; k_- \ll \omega. \tag{5.25}$$

Because the expression in (5.23) or (5.25) is invariant under Lorentz transformations in the z direction, it holds not only in the center-of-mass system but also in any system in which both incident particles have large momenta in opposite directions along the z-axis. Let the large plus momentum be ω_1 and the large minus momentum be ω_2; then equations (5.23) and (5.25) hold so long as

$$k_+ \ll \omega_1, \qquad k_- \ll \omega_2. \tag{5.26}$$

The integrated cross section for pair production is large when s is large. To see this, we integrate $d\sigma$ in equation (5.23) over k_z. We find that the integrated cross section is proportional to

$$\int_{-\omega}^{\omega} \frac{dk_z}{\sqrt{k_z^2 + \mathbf{k}_\perp^2 + \mu^2}} = \int d\eta \simeq \ln s, \tag{5.27}$$

which is infinite as $s \to \infty$. Note that $\ln s$ is the length of the rapidity space available for a produced particle.

We can similarly consider the process of the creation of two or more pionization pairs. The lowest (sixth-) order diagrams for the production of two pionization pairs are illustrated in figure 5.4. As before, both Δ_- and Δ'_+ are as small as ω^{-1}. Thus, referring to figure 5.4, we find that q must supply the plus momentum to the lower pair and the minus momentum to the upper pair:

$$k_{1-} + k_{2-} \simeq -q_-,$$

$$k_{3+} + k_{4+} \simeq q_+. \tag{5.28}$$

Now, for the production amplitude to be large, q should be close to the mass shell. Thus $q_+ q_-$ cannot be large. Together with the aforementioned fact that the two particles in a pair must have comparable longitudinal momenta, we conclude that $k_{1-} k_{3+}$ cannot be large, or, equivalently, k_{3+}/k_{1+} cannot be large. Thus we have, roughly,

$$k_{1+} \geqslant k_{3+}. \tag{5.29}$$

Therefore the longitudinal momenta of the pairs are approximately ordered according to their positions in the diagram.

In the region $k_{1+} \gg k_{3+}$, the amplitude corresponding to the diagrams in figure 5.4 simplifies and is approximately equal to

$$\frac{g^6 \omega^2 m^{-2} F(k_1, k_2, \Delta) F(k_3, k_4, q)}{(\Delta_\perp^2 + \lambda^2)(\Delta'^2_\perp + \lambda^2)(\mathbf{q}_\perp^2 + \lambda^2)}. \tag{5.30}$$

The integrated cross section for this process is proportional to

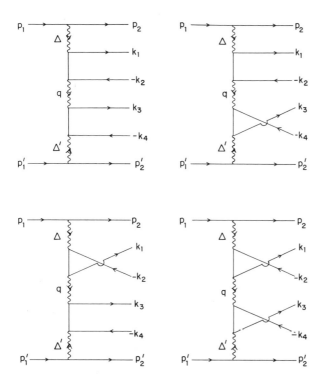

Figure 5.4
The lowest- (sixth-) order diagrams in which two created pairs can appear as pionization products.

$$\int d\eta_1 \int d\eta_3 = \tfrac{1}{2}\ln^2 s, \tag{5.31}$$

where η_1 and η_3 are the rapidities corresponding to k_1 and k_3, respectively. Note that the factor $\tfrac{1}{2}$ results from the restriction (5.29).

By repeating the same argument, we find that the lowest-order cross section for the production of n pairs is proportional to

$$\frac{1}{n!}\ln^n s. \tag{5.32}$$

Therefore, in the collision of two high-energy particles, the larger the number n of created pairs, the larger the corresponding production cross section. (This is true only if n is no larger than $\ln s$. Otherwise, the factor $(n!)^{-1}$ may become so small that expression (5.32) is small. Furthermore, our calculation is valid only in the limit $s \to \infty$ with n fixed.)

References

[1] H. Cheng and T. T. Wu, *Physical Review Letters* 23 (1969): 1311; R. P. Feynman, *Physical Review Letters* 23 (1969): 1415.

[2] J. Benecke, T. T. Chou, C. N. Yang, and E. Yen, *Physical Review* 188 (1969): 2159; T. T. Chou and C. N. Yang, *Physical Review Letters* 25 (1970): 1072; J. Benecke, *Nuovo Cimento A* 2 (1971): 615; J. P. Hsu, *Lettere al Nuovo Cimento* 3 (1970): 793, *Physical Review D* 3 (1971): 257; S. Y. Lo, *Lettere al Nuovo Cimento* 2 (1971): 1091.

[3] H. Cheng and T. T. Wu, *Physical Review D* 1 (1970): 456; L. N. Chang and N. P. Chang, *Physical Review D* 4 (1971): 1856; S. Tanaka, *Physical Review D* 4 (1971): 2419; W. R. Frazer, L. Ingber, C. H. Mehta, C. H. Poon, D. Silverman, K. Stowe, P. D. Ting, and H. J. Yesian, *Reviews of Modern Physics* 44 (1972): 284; T. Jaroszewicz and J. Wosiek, *Acta Physica Polonica B* 8 (1977): 837.

[4] S. J. Brodsky, T. Kinoshita, and H. Terazawa, *Physical Review Letters* 25 (1970): 972, *Physical Review D* 4 (1971): 1532; V. G. Serbo, *JETP Letters* 12 (1970): 39; K. Fujikawa, *Nuovo Cimento A* 12 (1972): 83, 117; S. K. Bose, *Lettere al Nuovo Cimento* 5 (1972): 662; H. Terazawa, *Reviews of Modern Physics* 45 (1973): 615; G. Bonneau and F. Martin, *Nuclear Physics B* 68 (1974): 367; G. Grammer and T. Kinoshita, *Nuclear Physics B* 80 (1974): 461.

[5] C. F. von Weizsäcker, *Zeitschrift für Physik* 88 (1934): 612; E. J. Williams, *Det Kongilege Danske Videnskabernes Selskab* 13 (1935): 4; F. E. Low, *Physical Review* 120 (1960): 582; N. Arteaga-Romero, A. Jaccarini, and P. Kessler, *Comptes Rendus*, Academy of Science (Paris), ser. B, 269 (1969): 153, 1129; H. Cheng and T. T. Wu, *Nuclear Physics B* 32 (1971): 461; C. E. Carlson and W. K. Tung, *Physical Review D* 6 (1972): 147; V. N. Baier and V. S. Fadin, *Soviet Physics JETP* 36 (1973): 399; C. J. Brown and D. H. Lyth, *Nuclear Physics B* 53 (1973): 323, 73 (1974): 417; G. Bonneau, M. Gourdin, and F. Martin, *Nuclear Physics B* 54 (1973): 573; V. L. Chernyak and V. G. Serbo, *Nuclear Physics B* 67 (1973): 464; V. M. Budnev, I. F. Ginzburg, G. V. Meledin, and V. G. Serbo, *Soviet Journal of Nuclear Physics* 16 (1973): 201; V. G. Serbo and V. L. Chernyak, *Soviet Journal of Nuclear Physics* 19 (1974): 66; I. F. Ginzburg, G. V. Meledin, and V. G. Serbo, *Soviet Journal of Nuclear Physics* 21 (1975): 409.

[6] A. H. Mueller, *Physical Review D* 2 (1970): 2963; C. L. Jen, K. Kang, P. Shen, and C. I. Tan *Physical Review Letters* 27 (1971): 458, *Annals of Physics* 72 (1972): 548; C. E. DeTar, *Physical Review D* 3 (1971): 128, 11 (1975): 866; N. F. Bali, A. Pignotti, and D. Steele, *Physical Review D* 3 (1971): 1167; D. Gordon and D. Veneziano, *Physical Review D* 3 (1971): 2116; V. V. Ezhela, A. A. Logunov, M. A. Mestvirishvily, V. A. Petrov, and G. L. Rcheulishvily, *Physics Letters B* 47 (1973): 501.

[7] D. E. Lyon, C. Risk, and D. M. Tow, *Physical Review D* 3 (1971): 104; H. J. Mück, M. Schachter, F. Selonke, B. Wessels, V. Blobel, A. Brandt, G. Drews, H. Fesefeldt, B. Hellwig, D. Mönkemeyer, P. Söding, G. W. Brandenburg, H. Franz, P. Freund, D. Lüers, W. Richter, and W. Schrankel, *Physics Letters B* 39 (1972): 303; V. V. Ezhela, A. A. Logunov, M. A. Mestvirishvily, et al., *Physics Letters B* 47 (1973): 501; C. O. Kim and K. S. Sim, *Physical Review D* 7 (1973): 2597; R. W. Brown, W. F. Hunt, K. O. Mikaelian, and I. J. Muzinich, *Physical Review D* 8 (1973): 3083.

6 Particle Production and the Elastic Amplitude

6.1 Introduction

Hadron physics and relativistic quantum field theory have a number of profound attributes in common. Some of them are relativistic kinematics, particle production, and unitarity.

In chapters 3 and 4, we showed that relativistic kinematics has much to do with the behavior of the elastic scattering. For example, we have seen that in the high-energy limit the phase shift in the semiclassical approximation approaches zero in nonrelativistic potential scattering but approaches a nonzero limit in relativistic potential scattering. Relativistic kinematics also plays an important role in determining the phase space region occupied by the produced particles. This is because the magnitude of an amplitude is affected by those of its propagators, which are crucially related to the kinematics.

To appreciate this, let us look at the role of kinematics in another example. Consider an incident particle of energy-momentum $(E, \mathbf{p})$ turning into two particles of energy-momentum $(E_1, \mathbf{p}_1)$ and $(E_2, \mathbf{p}_2)$, respectively. We use the conventional (pre-Feynman) perturbation method. Then the spatial momentum is conserved:

$$\mathbf{p} = \mathbf{p}_1 + \mathbf{p}_2, \tag{6.1}$$

and the propagator is equal to the inverse of the energy difference

$$(E - E_1 - E_2). \tag{6.2}$$

Therefore the amplitude is large if the energy difference (6.2) is small, subject to the constraint (6.1).

In relativistic kinematics, the energy and the magnitude of the spatial momentum of a high-energy particle are approximately equal, that is,

$$E \simeq |\mathbf{p}|,$$

and similarly for $(E_1, \mathbf{p}_1)$ and $(E_2, \mathbf{p}_2)$. It then follows that the energy difference (6.2) approximately vanishes if

$$|\mathbf{p}| = |\mathbf{p}_1| + |\mathbf{p}_2|. \tag{6.3}$$

Now, equation (6.1) means that $\mathbf{p}, \mathbf{p}_1$, and $\mathbf{p}_2$ form a triangle. Thus equations (6.1) and (6.3) can be satisfied together only if these three momenta are collinear. Therefore the momenta of the produced particles are approximately parallel to the incident momentum.

To be more quantitative, we choose $\mathbf{p}$ in the z direction and put $|\mathbf{p}| \equiv \omega$. We denote

$$\mathbf{p}_i = (x_i\omega, \mathbf{p}_{i\perp}), \qquad i = 1, 2, \tag{6.4}$$

where the first entity inside the parentheses is the z component. By equation (6.1), we have

$$\sum_{i=1}^{2} x_i = 1, \quad \text{and} \quad \sum_{i=1}^{2} \mathbf{p}_{i\perp} = 0. \tag{6.5}$$

In the limit $\omega \to \infty$, equation (6.3) is satisfied only if

$$|\mathbf{p}_{i\perp}| = O(1), \tag{6.6}$$

and

$$1 > x_i > 0, \qquad i = 1, 2. \tag{6.7}$$

More precisely, the energy difference (6.2) is as small as ω^{-1} if conditions (6.6) and (6.7) hold:

$$\sqrt{\omega^2 + \mu^2} - \sum_{i=1}^{2} \sqrt{x_i^2 \omega^2 + \mathbf{p}_{i\perp}^2 + \mu_i^2}$$

$$\simeq \left(\omega + \frac{\mu^2}{2\omega}\right) - \sum_{i=1}^{2} \left(x_i\omega + \frac{\mathbf{p}_{i\perp}^2 + \mu_i^2}{2x_i\omega}\right)$$

$$= \frac{1}{2\omega}\left(\mu^2 - \sum_{i=1}^{2} \frac{\mathbf{p}_{i\perp}^2 + \mu_i^2}{x_i}\right), \tag{6.8}$$

where μ and μ_i are the masses of the particles. This shows that the produced particles are not likely to possess large transverse momenta. Needless to say, this consideration can be trivially extended to the case of more than two produced particles.

The situation is entirely different in the case of nonrelativistic kinematics. In this case

$$E = \mathbf{p}^2/2m, \qquad E_i = \mathbf{p}_i^2/2m, \qquad i = 1, 2,$$

where we have chosen the mass of all of the particles to be m. Then the energy difference (6.2) vanishes if $\mathbf{p}$, $\mathbf{p}_1$, and $\mathbf{p}_2$ form a right triangle. Therefore, if nonrelativistic kinematics were valid, particles with large transverse momenta would be produced copiously.

Although relativistic kinematics dictates the favored phase space region, it does not tell us the magnitudes of the cross sections in the favored region. Indeed, in many relativistic field theories the production cross section at high energy vanishes in all regions. Gauge field theories stand as notable exceptions. In fact, we saw in chapter 5 that the lowest-order integrated cross section for the production of pairs is infinite as $s \to \infty$.

Such an abundance of particle production has a strong influence on the behavior of the elastic amplitude in gauge field theories. From the physical viewpoint this is a consequence of Huygen's principle. Consider a wave train impinging on a target. Let us consider this wave as a superposition of rays of various impact distances. If the ray at a certain impact distance produces particles, this ray is no longer in the initial state but is instead in an inelastic state. Thus it is lost to the incident beam. Huygen's principle states that the amplitude of the wave at a later time can be obtained by adding the contributions from all points of the wave train at any previous time. Thus, if the contribution from the ray at a certain impact distance is lost to an inelastic process, the wave at a later time is modified. In this way the inelastic processes can significantly affect the elastic scattering amplitude.

It follows from such considerations that the elastic amplitude and the inelastic amplitudes are related. One such relation is the unitarity condition for the S-matrix:

$$S^+ S = 1, \tag{6.9}$$

or

$$\sum_n S^*_{nf} S_{ni} = \delta_{fi}. \tag{6.10}$$

Let

$$S_{fi} = \delta_{fi} + (2\pi)^4 \delta^{(4)}(P_f - P_i) i \mathcal{M}_{fi} \Big/ \left(\prod_k f_k \right), \tag{6.11}$$

where f_k is the kinematic factor (see the first equation in section 3.3) for the kth external particle. Substituting equation (6.11) into equation (6.10), we get

$$\operatorname{Im} \mathcal{M}_{fi} = \frac{1}{2} \sum_n \frac{\mathcal{M}^*_{nf} \mathcal{M}_{ni}}{\prod_k f_k^2} (2\pi)^4 \delta^{(4)}(P_n - P_i), \tag{6.12}$$

where the index k runs over all the particles in the intermediate state n. If f and i are both two-fermion states, then $\mathcal{M}_{fi}$ is the elastic fermion-fermion scattering amplitude. Equation (6.12) says that the imaginary part of the elastic scattering amplitude is equal to a sum of contributions, each of which represents a scattering process of $i \to n \to f$. In particular, if we set $i = f$, the left-hand side of equation (6.12) is the imaginary part of the fermion-fermion scattering amplitude in the forward direction, whereas the right-hand side is the sum of the absolute value squared of all transition amplitudes and is hence proportional to the total cross sections. Taking into account the kinematic factors properly, we get

$$\text{Im}\, \mathcal{M}_{ii} = \frac{4qE}{m^2} \sigma_{\text{tot}}, \tag{6.13}$$

where q and E are the momentum and the energy, respectively, of an incoming fermion in the center-of-mass system. Equation (6.13) is known as the optical theorem.

6.2   The Tower Diagram

The considerations just discussed allow us to investigate the contribution of pair creation to the elastic scattering amplitude. In the lowest order the integrated cross section of one-pair creation over the pionization region increases as $\ln s$ as $s \to \infty$. Thus, by the optical theorem (6.13), the imaginary part of the elastic scattering amplitude in the forward direction resulting from this inelastic contribution asymptotically approaches $s \ln s$ times a constant. This is larger than the multiphoton exchange amplitude by a factor of $\ln s$ and renders the conjecture of constant total cross section suspect. Such an amplitude deserves closer study.

The lowest-order process of pionization is represented by the diagrams in figure 5.2. Thus the lowest-order contribution of the pionization process to the elastic amplitude is represented by the diagrams in figure 6.1. These diagrams describe the physical processes in which a pair is created and then annihilated, without appearing in the initial or the final state. Indeed, figures 6.1a and 6.1c have unitarity cuts (the dashed lines in the diagrams) that separate each of them into two diagrams in figure 5.2. We can therefore think of the corresponding process as $f + f \to f + f + f + \bar{f} \to f + f$. Diagram 6.1b is related to the other diagrams in figure 6.1 by gauge invariance and must be included.

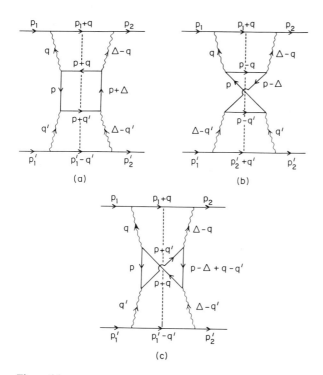

Figure 6.1
The lowest- (eighth-) order diagrams for fermion-fermion elastic scattering that give amplitudes as large as $s \ln s$.

The calculations for these diagrams are straightforward but tedious, and we do not present them until chapter 11. The result is

$$\mathcal{M}_1 \simeq \frac{ig^4 s \ln s}{4m^2} \int \frac{d^2q_\perp d^2q'_\perp}{(2\pi)^4}$$

$$\times \frac{K_t(\mathbf{q}_\perp, \mathbf{q}'_\perp, \Delta)}{(\mathbf{q}_\perp^2 + \lambda^2)(\mathbf{q}'^2_\perp + \lambda^2)[(\Delta - \mathbf{q}_\perp)^2 + \lambda^2][(\Delta - \mathbf{q}'_\perp)^2 + \lambda^2]}, \quad (6.14)$$

where

$$K_t(\mathbf{q}_\perp, \mathbf{q}'_\perp, \Delta) = \frac{g^4}{(2\pi)^4} \int d^2p_\perp \int_0^1 \frac{dx}{x\mathbf{p}_\perp^2 + (1-x)(\mathbf{p}_\perp + \mathbf{q}_\perp)^2 + m^2}$$

$$\times \left\{ \frac{\mathrm{Tr}[(\not{p}_\perp + m)(-\not{p}_\perp - \not\Delta + m)][(\mathbf{p}_\perp + \mathbf{q}_\perp)^2 + m^2]}{x(\mathbf{p}_\perp + \Delta_\perp)^2 + (1-x)(\mathbf{p}_\perp + \mathbf{q}_\perp)^2 + m^2} \right.$$

$$\left. - \frac{\mathrm{Tr}[(\not{p}_\perp + m)(-\not{p}_\perp - \not{q}'_\perp + m)(\not{p}_\perp - \not\Delta + \not{q}_\perp + \not{q}'_\perp + m)(-\not{p}_\perp - \not{q}_\perp + m)]}{x(\mathbf{p}_\perp + \mathbf{q}'_\perp)^2 + (1-x)(\mathbf{p}_\perp - \Delta + \mathbf{q}_\perp + \mathbf{q}'_\perp)^2 + m^2} \right\}. $$
$$(6.15)$$

We note that the amplitude in relation (6.14) is imaginary and is of the order $s \ln s$ for all t. At $t = 0$, this result agrees with the result obtained earlier by using the optical theorem.

It is convenient to express the amplitude (6.14) in simplified notation. Let

$$\mathcal{K}_t(\mathbf{q}_{1\perp}, \mathbf{q}_{2\perp}, \Delta) \equiv K_t(\mathbf{q}_{1\perp}, \mathbf{q}_{2\perp}, \Delta) \bigg/ \prod_{i=1}^2 (\mathbf{q}_{i\perp}^2 + \lambda^2)^{1/2}[(\Delta - \mathbf{q}_{i\perp})^2 + \lambda^2]^{1/2},$$
$$(6.16)$$

and define the corresponding operator $\mathcal{K}_t$ by

$$(\mathcal{K}_t F)(\mathbf{q}_\perp, \Delta) \equiv \int \frac{d^2q_\perp}{(2\pi)^2} \mathcal{K}_t(\mathbf{q}_\perp, \mathbf{q}'_\perp, \Delta) F(\mathbf{q}'_\perp, \Delta). \quad (6.17)$$

We also define the scalar product

$$(F_1, F_2) \equiv \int \frac{d^2q_\perp}{(2\pi)^2} F_1(\mathbf{q}_\perp, \Delta) F_2(\mathbf{q}_\perp, \Delta), \quad (6.18)$$

and

$$J(\mathbf{q}_\perp, \Delta) \equiv \frac{g^2}{2m} \bigg/ \prod_{i=1}^2 (\mathbf{q}_{i\perp}^2 + \lambda^2)^{1/2}[(\Delta - \mathbf{q}_{i\perp})^2 + \lambda^2]^{1/2}. \quad (6.19)$$

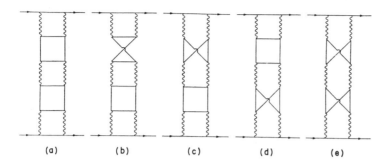

Figure 6.2
The lowest- (twelfth-) order diagrams for fermion-fermion elastic scattering that give amplitudes as large as $s(\ln s)^2$.

In this notation the amplitude in relation (6.14) is simply

$$\mathcal{M}_1 \simeq is \ln s (J, \mathcal{K}_t J). \tag{6.20}$$

Because we have found that there is a term as large as $s \ln s$ in the elastic amplitude, it is natural to ask if there are terms even larger, say $s \ln^2 s$. The answer is yes. The $s \ln^2 s$ terms first appear in diagrams of two-pair creation, as the lowest-order cross section for two-pair creation is proportional to $\ln^2 s$. Specifically, the relevant Feynman diagrams are those illustrated in figure 6.2. More generally, the lowest-order Feynman diagrams that give $s \ln^n s$ terms, $n = 1, 2, \ldots, n$, are the ones with n electron loops joined together vertically, with each loop having four vertices. We call such diagrams *tower diagrams*. The calculations for these diagrams of arbitrary n are given in chapter 11. The result is

$$\mathcal{M}_n \simeq \frac{is(\ln s)^n}{n!}(J, \mathcal{K}_t^n J), \tag{6.21}$$

where $\mathcal{M}_n$ is the amplitude for the tower diagrams with n loops. Summing over n, we get

$$\mathcal{M}_T \equiv \prod_{n=0}^{\infty} \mathcal{M}_n \simeq is(J, s^{\mathcal{K}_t} J). \tag{6.22}$$

The subscript t in $\mathcal{K}_t$ stands for tower.

It follows from equation (6.22) that the asymptotic behavior of $\mathcal{M}_T$ is determined by the largest eigenvalue of $\mathcal{K}_t$. As we show in Appendix B, this value is equal to $11\alpha^2\pi/32$, and $\mathcal{M}_T$ is asymptotically proportional to [1]

$$s^{1+11\alpha^2\pi/32}(\ln s)^{-3/2}. \tag{6.23}$$

The factor $(\ln s)^{-3/2}$ is due to the fact that this largest eigenvalue is not a discrete eigenvalue but is at the endpoint of a continuous spectrum. In the language of Regge poles, the leading singularity in the J-plane for $\mathcal{M}_T$ is a branch point, not a pole, and is located at $J > 1$, not at $J = 1$.

The diagrams in figure 6.1 are the lowest-order ones that give terms as large as $s \ln s$. Thus the amplitude given by relation (6.20) is the largest amplitude (the leading term) of the eighth order. Similarly, the amplitude (6.21) is the largest amplitude (the leading term) of the $4(n + 1)$th order. Thus $\mathcal{M}_T$ given by equation (6.22) is equal to the sum of the leading terms. One might hope that, by summing the leading term of each order, one might obtain some semblance of the asymptotic form of the scattering amplitude in the high-energy limit. The mystery of high-energy scattering turns out to be more elusive. Indeed, it has been proved by Froissart [2] that the elastic scattering amplitude in the forward direction cannot be larger than $s \ln^2 s$ in the limit $s \to \infty$. Because amplitude (6.23) is larger than the Froissart bound, it cannot be the asymptotic form of the elastic amplitude. As we will see in the next two chapters, the interpretation of expression (6.23) leads to dramatic and far-reaching consequences that are entirely unexpected.

References

[1] H. Cheng and T. T. Wu, *Physical Review Letters* 24 (1970):759, *Physical Review D* 1 (1970):2775; G. V. Frolov, V. N. Gribov, and L. N. Lipatov, *Physics Letters B* 31 (1970):34; W. J. Pardee, *Physical Review D* 3 (1971):1655; J. L. Cardy, *Physics Letters B* 53 (1974):355, *Nuclear Physics B* 93 (1975):525.

[2] M. Froissart, *Physical Review* 123 (1961):1053; A. Martin, *Nuovo Cimento* 42 (1966):930, 44 (1966):1219.

7 The Theory of Expanding Protons

Yu Gung (foolish old man) lived near the ocean. Unfortunately, there was a mountain blocking his ocean view. One day, Yu Gung decided to remove the cursed mountain and started digging away with his sons. A wise old man passed by and exclaimed: "Fools! You can never get it done this way." Yu Gung replied: "A mountain cannot grow bigger. Therefore, our digging must be making it smaller. Besides, I enjoy digging very much."

So, Yu Gung persisted, day after day and year after year. One morning Yu Gung rose and found the mountain gone. It had collapsed and fallen into the ocean during a storm.

Fable from ancient China

7.1 The Rising Total Cross Section

We are now in a position to describe the reasoning that leads to the theory of expanding protons, which predicted the rise of the total cross sections [1]. From the late 1950s to the early 1970s, hadronic scattering at high energies and small angles was one of the most intensely studied topics in elementary particle physics. On the experimental side, many such processes, both elastic and inelastic, were measured. On the theoretical side, a number of useful models were invented. Although these models covered a wide spectrum, most of them can be classified as belonging to one or the other of two general schools of thought: one emphasizing s-channel unitarity, such as the droplet model of Yang and collaborators [2], and the other emphasizing t-channel unitarity, such as the Regge model [3]. Although both types of model were well motivated and had experimental support, they nearly directly contradicted each other.

Despite all their differences, the various theories shared one cornerstone on which there was almost unanimous agreement: The total cross section approaches a constant in the limit of infinite energy. Until the early 1970s, there was both experimental and theoretical support for this assumption. On the experimental side, the measured total cross sections for pp, $\pi^+ p$, $\pi^- p$, as well as $K^- p$ appeared to approach constant values at laboratory energies of 20 GeV or so. Moreover, the results from Serpukhov between 20 GeV and 70 GeV showed almost no feature whatsoever. On the theoretical side, this assumption of constant total cross sections agreed with a well-established feature of wave motion—that at high frequencies the propagation of a wave can be described semiclassically and approaches the limit of geometrical optics. Further support can be found in quantum mechanics, as in the Dirac equation [4] and the Klein-Gordon equation

[5]. Indeed, the evidence was considered to be so strong that many tended to forget that it was an assumption. For example, Pomeranchuk, one of the pioneers of high-energy scattering, assured us at the beginning of a celebrated paper [6] that this result "follows simply from the fact that all strong interactions approach zero exponentially for large values of the impact parameter."

In the late 1960s many people, including us, attempted to seek clues from quantum gauge field theories. Although it was not immediately recognized, the major conclusion from such studies is unavoidable: The total cross section increases indefinitely as the energy increases [1]. This means that, theoretically, a proton can become as big as a house and that it interacts strongly with every hadron in its path! As expected, such a seemingly ridiculous conclusion was not immediately accepted. However, after three years, in 1973 this theoretical prediction of rising total cross sections was beautifully verified by two experimental groups, a Pisa–Stony Brook collaboration [7] and a CERN-Rome collaboration [8], both working at the Intersecting Storage Rings (ISR) at CERN. These experimental groups found a dramatic upturn of the proton-proton total cross section at equivalent laboratory momenta between 291 and 1480 GeV/c. Furthermore, this experimentally observed rise agreed reasonably well with the value obtained previously from a simple phenomenological model based on the results from quantum gauge field theory [9]. Gauge field theory had scored another theoretical triumph.

7.2   Physical Picture of the Theory of Expanding Protons

In retrospect, the theory of expanding protons is a natural consequence of gauge field theories. The intuitive picture is as follows.

1. First, the spin of the gauge boson is necessarily 1 for both Abelian and non-Abelian gauge theories. From the study of Regge poles we have learned that the amplitude for the exchange of a particle of spin J is proportional to

$$s^J \tag{7.1}$$

in the high-energy limit $s \to \infty$. For $J = 1$, this amplitude is proportional to s. More generally, the amplitude for the exchange of n particles of spins $J_1, J_2, \ldots, J_n$ is proportional to

$$s^{J_1 + J_2 + \cdots + J_n - n + 1} \tag{7.2}$$

in this high-energy limit. If each of the J's is equal to 1, then expression (7.2) gives again s, as exemplified by the multi-vector-meson exchange studied explicitly in chapter 3. After dividing by the kinematic factors, we find that the amplitude proportional to s leads to a total cross section independent of s in the high-energy limit.

On the other hand, the cross section resulting from the exchange of scalar mesons ($J = 0$), for example, vanishes as s^{-2}. Thus the slow variation of the total cross section with the energy observed experimentally is explained by a gauge boson mediating strong interactions.

2. The total cross section, although slowly varying, is strictly not a constant. Equivalently, the elastic scattering amplitude divided by s is not a constant. This behavior is due to the existence of production processes. As was discussed in section 5.3 the lowest-order cross section for the production of n pairs is proportional to $(\ln s)^n$. Thus the total cross section has logarithmic factors of s. By the s-channel unitarity condition, the elastic scattering amplitude has terms proportional to $s (\ln s)^n$, $n = 1, 2, \ldots$.

3. The factor $\ln s$ is the size of the rapidity space kinematically allowed for a created particle. As s increases, this size increases. Furthermore, n (the number of created pairs kinematically allowed) also increases with s. Thus the sum over n of the lowest-order production cross sections increases faster than any power of $\ln s$. For the elastic scattering amplitude, the corresponding sum is similar to that encountered in Regge models. Indeed, this sum over n is asymptotically proportional to

$$s^{1+a}(\ln s)^{-3/2} \tag{7.3}$$

with $a > 0$. The crucial feature of expression (7.3) is with the power of s. This power is independent of t, unlike the behavior of a Regge pole. More important, this power is always larger than 1.

4. The large amplitude of expression (7.3) is *not* to be interpreted as a violation of the Froissart bound or of the unitarity condition. Rather, it is a manifestation of the strongly absorptive interaction between two high-energy colliding particles. The sum of tower amplitudes, having a large *imaginary* part, can be regarded as representing a strongly absorptive potential, as conceived in Froissart's original paper [10]. It means that the transition probability of particle production increases as a power of s. Now, the incident plane wave is a superposition of waves with impact distance b ranging from zero to infinity. If b is small, the wave is prone to creating slow particles and is *lost to the beam*. It may be worth emphasizing that,

for a particle to be lost to the beam, it is not necessary for anything dramatic to happen. For example, consider the scattering of a 1000-GeV proton with a 1000-GeV antiproton. Each of the two incident particles may lose 140 MeV to create a pair of charged pions, with the resulting energies of 999.86 GeV each. Because of beam spread, such a small change of energy cannot be detected even experimentally. Nevertheless, these particles are considered to be lost to the beam. On the other hand, if b is large, the wave is unlikely to interact with the target and may therefore survive. What is the value of b at which the attenuation of the incident wave is still appreciable? The absorption is controlled by the Yukawa tail [11], roughly of the form $e^{-\mu b}$ (where μ is a constant) decreasing exponentially with b. Because the cross section for creating pionization pairs increases as s^a, attenuation remains appreciable if the product of these two factors is of the order of 1,

$$s^a e^{-\mu b} = O(1),$$

or b is of the order of R given by

$$R = \frac{a}{\mu} \ln s. \tag{7.4}$$

Because R increases logarithmically with the energy, the total cross section must increase as $(\ln s)^2$ in the limit of high energies, and the imaginary part of the elastic scattering amplitude increases as $s(\ln s)^2$. This is exactly the Froissart bound, and it is consistent with unitarity.

These considerations lead to a dramatic physical picture for high-energy hadron-hadron scattering. At extremely high energies, a particle acts like a Lorentz-contracted pancake that can be roughly separated into the following two regions: (1) a black core (completely absorptive) with a radius expanding logarithmically with energy and (2) a gray fringe (partially absorptive) that extends farther out. The width of this gray fringe can be estimated by setting

$$s^a e^{-\mu b} = C,$$

or

$$b = \frac{a}{\mu} \ln s - \frac{\ln C}{\mu}.$$

As C varies from $\frac{1}{2}$ to 1, for example, b changes by an amount $(\ln 2)/\mu$. Thus the width of the gray fringe is independent of s. This physical picture in the

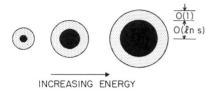

$\overline{O(1)}$
$O(\ell n\ s)$

INCREASING ENERGY

Figure 7.1
Representation of the appearance of a high-energy particle in the theory of expanding protons.

theory of expanding protons is illustrated in figure 7.1. It is different from that in the Regge pole model and from that in the droplet model. In each of these two models a particle is pictured either as a sphere or a disk, depending on the Lorentz frame used. In the Regge pole model the disk becomes larger and more transparent as energy increases; in the droplet model both the size and the opacity of the disk are independent of the energy.

7.3   Further Consequences of the Theory of Expanding Protons

The asymptotic form of the elastic scattering amplitude can be obtained by applying Huygen's principle. This principle states that, if the amplitude of the wave is known on a given surface, the amplitude at a nearby point is equal to the sum of contributions from all points of the surface. In a crude approximation we can consider the wave to be entirely absorbed if $b < R$ and entirely unchanged if $b > R$. Thus the elastic scattering amplitude is equal to

$$\mathscr{M} \simeq \frac{is}{2m^2} \int d^2 b e^{-i\Delta \cdot \mathbf{b}} [1 - S(\mathbf{b}, s)], \tag{7.5}$$

where

$$S(\mathbf{b}, s) \simeq \begin{cases} 0, & b < R, \\ 1, & b > R \end{cases} \tag{7.6}$$

and where $s/2m^2$ is a kinematic factor. Relation (7.5) states that the scattering of a high-energy particle from a target particle is like that from a black (completely absorbing) disk with radius R increasing logarithmically with the energy. With the approximation (7.6), the integral in relation (7.5) can be explicitly carried out, and we obtain

$$\mathcal{M} \simeq \frac{is\pi}{m^2} \int_0^R b \, db \, J_0(\Delta b)$$

$$= \frac{i\pi s R}{m^2 \Delta} J_1(\Delta R). \tag{7.7}$$

By the optical theorem, the total cross section is related to the imaginary part of the elastic scattering amplitude in the forward direction. Thus we get from relation (7.5) that

$$\sigma_{\text{tot}} \simeq 2 \int d^2 b \, \text{Re}[1 - S(\mathbf{b}, s)] = 2\pi R^2. \tag{7.8}$$

Therefore the total cross section is approximately equal to twice the area of the black disk where strong absorption takes place.

From relation (7.7), the usual diffraction peaks are expected for the elastic scattering amplitude around the forward direction. Indeed, let β_n be the roots of $J_1(\beta\pi)$; then the zeros of the elastic scattering amplitude are given by

$$\Delta R = \beta_n \pi, \qquad n = 1, 2, \ldots .$$

In particular, because $\beta_1 = 1.2197$, we have

$$\Gamma \sigma_{\text{tot}} = 2\pi^3 \beta_1^2 + O[(\ln s)^{-1}]$$

$$\approx 35.92 \text{ mb (GeV}/c)^2, \tag{7.9}$$

independent of the processes, where Γ is the value of Δ^2 at which the first dip occurs. Note also that the width of the diffraction peak shrinks as s increases. Relation (7.7) also implies that the integrated elastic cross section σ_{el} is given by

$$\sigma_{\text{el}} = \pi R^2 + O(\ln s), \tag{7.10}$$

and

$$\sigma_{\text{el}}/\sigma_{\text{tot}} = \tfrac{1}{2} + O[(\ln s)^{-1}]. \tag{7.11}$$

Furthermore, it follows from the dispersion relation that

$$\left. \frac{\text{Re } \mathcal{M}}{\text{Im } \mathcal{M}} \right|_{t=0} = \frac{\pi}{\ln s} + O[(\ln s)^{-2}], \tag{7.12}$$

which approaches zero from above as $s \to \infty$.

We can generalize relation (7.5) to an inelastic process in which one or both of the outgoing particles are excited states of the incoming particles. This can be done by regarding $(1 - S)$ in relation (7.5) as a matrix. The matrix element of the unity matrix 1 corresponding to an inelastic process is 0. Thus the matrix element of $(1 - S)$ corresponding to an inelastic process is appreciable only in the gray fringe. Consequently, both the forward amplitude and the integrated cross section for a process of diffractive excitation increase as $\ln s$. This result also holds for other processes, such as the process of pair creation.

Although there remain mathematical difficulties in calculating the high-energy scattering amplitudes more precisely, attenuation in the presence of strongly absorptive processes is a universal phenomenon occurring in classical physics and quantum physics alike. Readers interested in an in-depth study of high-energy scattering are referred to chapter 13, where a more extensive treatment is given for amplitudes in both QED and QCD to all orders.

References

[1] H. Cheng and T. T. Wu, *Physical Review Letters* 24 (1970): 1456.

[2] N. Byers and C. N. Yang, *Physical Review* 142 (1966): 976; T. T. Chou and C. N. Yang, *Physical Review Letters* 20 (1968): 1213. See also reference [1] of chapter 4.

[3] T. Regge, *Nuovo Cimento* 14 (1959): 951, 18 (1960): 947. See also references [2, 5] of chapter 4.

[4] P. A. M. Dirac, *Proceedings of the Royal Society of London* 117 (1928): 610, 118 (1928): 351.

[5] E. Schrödinger, *Annalen der Physik* 81 (1926): 109; W. Gordon, *Zeitschrift für Physik* 40 (1926): 117; O. Klein, *Zeitschrift für Physik* 41 (1927): 407.

[6] I. Ya. Pomeranchuk, *Soviet Physics JETP* 7 (1958): 499.

[7] S. R. Amendolia, G. Bellettini, P. L. Braccini, C. Bradaschia, R. Castaldi, V. Cavasinni, C. Cerri, T. Del Prete, L. Foa, P. Giromini, P. Laurelli, A. Menzione, L. Ristori, G. Sanguinetti, M. Valdata, G. Finocchiaro, P. Grannis, D. Green, R. Mustard, and R. Thun, *Physics Letters B* 44 (1973): 119.

[8] U. Amaldi, R. Biancastelli, C. Bosio, G. Matthiae, J. V. Allaby, W. Bartel, G. Cocconi, A. N. Diddens, R. W. Dobinson, and A. M. Wetherell, *Physics Letters B* 44 (1973): 112.

[9] H. Cheng, J. K. Walker, and T. T. Wu, "Impact picture of very high-energy hadron interactions (with numerical results)," Paper 524, 16th International Conference on High Energy Physics, 1972 (unpublished); T. T. Wu and H. Cheng, in *High Energy Collisions*, C. Quigg. ed. (New York: American Institute of Physics, 1973), 54.

[10] M. Froissart, *Physical Review* 123 (1961): 1053.

[11] H. Yukawa, *Proceedings of the Physico-Mathematical Society of Japan*, 17 (1935): 48.

8 Phenomenology

8.1 Introduction

In the preceding chapter the theoretical prediction of the rising total cross section was discussed in a model-independent way. When this general prediction was first arrived at in 1970 [1], there was no experimental indication whatsoever of this new phenomenon. In order to make the theoretical prediction more quantitative and hence useful to the experimentalists, it was necessary to develop phenomenological models that incorporate information about the structures of the various hadrons. Such information was necessary in order to differentiate between, for example, pp and πp scattering, and could be obtained from relatively low-energy data. In this chapter these phenomenological models are described together with their predictions on the total cross section and elastic differential cross section. Recently, some of these predictions for $p\bar{p}$ scattering have been found to be in excellent agreement [2, 3] with the CERN $p\bar{p}$ Collider data taken a decade later [4, 5].

It is by now generally accepted that strong interactions are described by quantum chromodynamics (QCD). This belief is based mostly on hard processes, that is, processes involving large momentum transfers. For example, some of the most direct evidence for the quark is from the two-jet events in electron-positron annihilation, first observed at SPEAR and interpreted as [6]

$$e^- e^+ \to q\bar{q}. \tag{8.1}$$

The gluon was discovered experimentally [7] and its spin measured [8] at PETRA through the three-jet events interpreted as

$$e^- e^+ \to q\bar{q}g. \tag{8.2}$$

The angular distributions for both the two-jet and the three-jet events are in good agreement with the results of low-order perturbative QCD (see chapter 1).

QCD is asymptotically free [9]. This means that the effective quark-gluon coupling constant is small for large momentum transfers but large for small momentum transfers. For this reason quantities controlled by small momentum transfers, such as the total cross sections, are not adequately described by low-order perturbative QCD. Instead, a systematic study of higher-order effects, as presented in the preceding chapters, is needed in order to understand these physical phenomena.

8.2 Models for Elastic Scattering

In this chapter three phenomenological models are described. They are of course closely related, each one being an improvement on the earlier one, rendered possible by the availability of better experimental data. Before there were any data to indicate the rise of total cross sections, model 0 was proposed [10] in order to give a semiquantitative estimate of the energy needed and the amounts of increase. In the following year the first data on the increase of proton-proton total cross section were obtained at the CERN ISR (Intersecting Storage Rings) simultaneously by the Pisa–Stony Brook Collaboration [11] and the CERN-Rome Collaboration [12]. Compared with this data, model 0 turned out to be fairly accurate, underestimating the increase by about 35%. By incorporating these CERN experimental data into the basic idea of this original model 0, we obtained model 1 [13], which is the first quantitatively accurate phenomenological model. Many predictions [14] have been obtained on the basis of this model 1, one example being the energy dependence of the ratio of the real to imaginary parts of the proton-proton elastic scattering amplitude in the forward direction [15]. Although all measurements at that time gave a negative value for this ratio, the theoretical prediction was that this ratio must, with increasing energies, become positive and then turn around to approach zero slowly. Shortly thereafter, this prediction was accurately verified by experimental data from Fermilab [16]. A more striking experimental verification, however, occurred much later, after the completion of the CERN proton-antiproton collider. At a center-of-mass energy of 546 GeV, not only the $p\bar{p}$ total cross section but also the $p\bar{p}$ elastic differential cross section in the range $|t| = 0.05$ to 0.25 $(GeV)^2$ agreed almost perfectly with those of model 1 given nine years earlier [4, 5, 14]. More recently, using all the pp and $p\bar{p}$ elastic data from the CERN ISR and the $p\bar{p}$ data from the CERN $p\bar{p}$ Collider, an overall fit has been carried out [3]. This model 2 is an improvement over model 1 for larger momentum transfers. Predictions for center-of-mass energies up to the range of nearly 100 TeV have been obtained on the basis of this model 2. Experimental verifications in this high-energy end, however, will not be forthcoming until the completion of the SSC (Superconducting Super Collider) or the CERN LHC (Large Hadron Collider).

For definiteness, we concentrate mainly on elastic pp and $p\bar{p}$ scattering. The guiding principle is to develop the *simplest* models consistent with the data and with the theoretical results of the last chapter. Before going into the details, we discuss first a point of paramount importance. In obtaining the logarithmically increasing radius of the black core as discussed in chapter 7, we must have, at least roughly, a Yukawa interaction at large impact distances. More precisely, the important factor is the $e^{-\lambda r}$ in the Yukawa form $e^{-\lambda r}/r$, and it reflects the absence of a massless hadron. It is thus necessary to incorporate in the model a dependence of the form $e^{-\lambda r}$ for large r; this factor cannot be replaced by either a power r^{-n} or a Gaussian e^{-r^2}.

The difference between pp elastic scattering and $p\bar{p}$ elastic scattering is due to Regge exchanges [17]. In the simple models to be described here, these Regge exchanges are included only for the total cross sections (see equations (8.10)), not for the differential cross sections. In other words, the pp and $p\bar{p}$ elastic amplitudes are approximated by the *same* impact-distance representation (see, however, section 2 of [8]):

$$\mathcal{M}(s, \Delta) = \frac{is}{2\pi} \int d\mathbf{x}_\perp e^{-i\Delta \cdot \mathbf{x}_\perp} D(s, \mathbf{x}_\perp), \tag{8.3}$$

where Δ is the momentum transfer and all spin variables have been omitted. The simplest form that we can use for the opacity $D(s, \mathbf{x}_\perp)$ is

$$D(s, \mathbf{x}_\perp) = 1 - e^{-\Omega(s, x_\perp^2)}, \tag{8.4}$$

with

$$\Omega(s, x_\perp^2) = S(s)F(x_\perp^2), \tag{8.5}$$

where $S(s)$ is a function of s. This choice of equation (8.5) includes in particular the assumption of factorization into a function of s and a function of $x_\perp^2$, so that we need to choose two functions of one variable rather than a function of two variables. This and other simplifying assumptions are discussed in section 8.6.

The following point should be noted. Suppose that we are interested instead in $\pi^+ p$ and $\pi^- p$ elastic scattering at high energies. In these cases the total cross sections are much smaller, and hence a different $F(x_\perp^2)$ must be used. In contrast, because $S(s)$ characterizes the Pomeron, it does *not* depend on the elastic process under consideration. In other words, the *same* $S(s)$ must be used for $\pi^\pm p$ and $K^\pm p$ as for pp and $p\bar{p}$.

How do we choose this universal function $S(s)$? Because pp elastic scattering is related to $p\bar{p}$ elastic scattering by crossing symmetry and because equation (8.4) applies equally well to both processes, $S(s)$ must be invariant under crossing. From the theoretical result (6.23), the simplest choice is

$$S(s) = \frac{s^c}{(\ln s)^{c'}} + \frac{u^c}{(\ln u)^{c'}},\tag{8.6}$$

where u is the third Mandelstam variable and c and c' are two constants to be determined. Note that $S(s)$ is complex.

A word must be said about the choice of units. In an asymptotic formula for large s, it is acceptable to write $\ln s$, because $\ln(s/s_1)$ and $\ln(s/s_2)$ for two different s_1 and s_2 differ by $O(1)$, which is smaller than either. On the other hand, because equation (8.6) is to be used in the phenomenological models, its right-hand side has no meaning without a definite choice of units. Fortunately, as discussed further in the next section, the results are not sensitive to this choice. In all published papers in which equation (8.6) is used, the unit for $\sqrt{s}$ is 1 GeV, nearly the proton mass.

The next task is to choose the function $F(x_\perp^2)$. In this case there is less guidance. From the discussion at the beginning of this section, we want

$$F(x_\perp^2) \sim \text{const } e^{-\lambda x_\perp}\qquad(\text{large } x_\perp).$$

Furthermore, because $x_\perp = 0$ merely means that the center of mass of the three quarks or antiquarks in the proton or antiproton are lined up and hence does not imply any head-on quark-quark or quark-antiquark scattering, $F(x_\perp^2)$ should be chosen to be smooth at $x_\perp^2 = 0$. (Note that, as a function of the two components of $\mathbf{x}_\perp$, $e^{-\lambda x_\perp}$ has a cusp at the origin.) The simplest choice consistent with both of these requirements is

$$F(x_\perp^2) = f \exp[-\lambda(x_\perp^2 + x_0^2)^{1/2}].\tag{8.7a}$$

This choice is used in both model 0 and model 1. For model 2 this $F(x_\perp^2)$ is instead taken to be related to the electromagnetic form factor $G(t)$ of the proton. More precisely, the Fourier transform of $F(x_\perp^2)$ is the product of $[G(t)]^2$ with a slowly varying function of t. Such a slowly varying factor should be present theoretically (see section 11.4) and is also needed phenomenologically for momentum transfers near the first shoulder [18]. The specific choice used is

$$\tilde{F}(t) = f[G(t)]^2 \frac{a^2 + t}{a^2 - t}.\tag{8.7b}$$

Equations (8.3) through (8.7) specify completely the elastic amplitudes for pp and $p\bar{p}$ scattering, taken to be the same by neglecting Regge exchanges. In terms of this $\mathcal{M}(s, \Delta)$, the differential cross section is given by

$$\frac{d\sigma}{dt} = \pi |s^{-1}\mathcal{M}(s, \Delta)|^2, \tag{8.8}$$

where $t = -\Delta^2$. The total elastic cross section is then, of course,

$$\sigma_{\mathrm{el}} = \int_0^\infty \frac{d\sigma}{dt} d|t|. \tag{8.9}$$

For the total cross section the desire to fit the relatively low-energy data implies that the effects of Regge exchanges cannot be entirely neglected. If the $\mathcal{M}(s, \Delta)$ of equation (8.3) is taken literally as the elastic scattering amplitude, then by the optical theorem the total cross section must be proportional to $\mathrm{Im}\,\mathcal{M}(s, 0)$. The inclusion of the Regge exchanges of ρ, ω, ϕ, etc., in the most simple-minded way leads to the following modification:

$$\sigma_{\mathrm{tot}}(pp) = A(pp)s^{-1/2} + \frac{4\pi}{s}\mathrm{Im}\,\mathcal{M}(s, 0),$$

$$\sigma_{\mathrm{tot}}(p\bar{p}) = A(p\bar{p})s^{-1/2} + \frac{4\pi}{s}\mathrm{Im}\,\mathcal{M}(s, 0). \tag{8.10}$$

(Remember that to convert from $(\mathrm{GeV})^{-2}$ to mb, a factor of $4.893/4\pi$ is needed.)

In order to satisfy the optical theorem, these two parameters $A(pp)$ and $A(p\bar{p})$ should also appear in the formula for $d\sigma/dt$. This requires a more complete treatment of Regge exchange; such a treatment has been carried out [18]. However, because the contributions of the $A(pp)$ and $A(p\bar{p})$ terms to the right-hand sides of equations (8.10) are relatively small in the energy ranges suitable for the phenomenological predictions, we use here equations (8.10) together with equation (8.8) in order to stay away from the more complicated details.

We conclude this section with a list of all the parameters to be used for pp and $p\bar{p}$ scattering. There are seven, all real, that can be conveniently classified:

1. λ, c, and c';

2. f and x_0 (or a); and

3. $A(pp)$ and $A(p\bar{p})$.

The three parameters in class 1 are universal in the sense that they apply to all hadronic processes, both elastic and inelastic diffraction scattering. They are most conveniently determined from pp elastic scattering. By contrast, the two parameters in class 2 depend on the elastic scattering under consideration, although they do not change under crossing. The two parameters in class 3 also depend on the process under consideration and, furthermore, are not invariant under crossing.

In order to make this point clear, suppose that we are interested simultaneously in the following six elastic scatterings in the context of model 1:

$$pp, \quad p\bar{p}, \quad \pi^+ p, \quad \pi^- p, \quad K^+ p, \quad K^- p.$$

Instead of seven parameters, we need fifteen:

1. λ, c, and c';
2. $f(pp)$, $x_0(pp)$, $f(\pi p)$, $x_0(\pi p)$, $f(Kp)$, and $x_0(Kp)$; and
3. $A(pp)$, $A(p\bar{p})$, $A(\pi^+ p)$, $A(\pi^- p)$, $A(K^+ p)$, and $A(K^- p)$.

Note that the six parameters of class 3 are used to take care of Regge background.

8.3   Determination of Phenomenological Parameters

The determination of the seven phenomenological parameters used in the models for pp and $p\bar{p}$ elastic scattering at high energies was carried out somewhat differently for models 0, 1, and 2.

During the construction of model 0, which began shortly after the theoretical prediction of increasing total cross section described in the preceding chapter, there was no experimental indication of this new phenomenon. The most important question at that stage was, What prevents c from being zero or small?

Because the increase in total cross sections is more sensitive to c than to c', this uncertainty in c made it moot to discuss the value of c'. Accordingly, c' is simply set to 0. At this first stage, the value of c was determined from the conditions that

$$A(K^+ p) = 0 \tag{8.11}$$

borrowed from duality and that the $K^+ p$ channel is exotic.

The result of this rough model is shown in figure 8.1. From this early

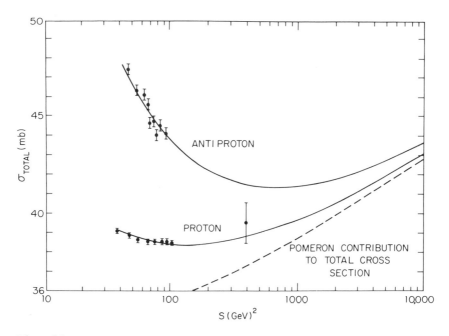

Figure 8.1
An early rough prediction of rising pp and $p\bar{p}$ total cross sections from model 0.

result the increase of the total pp cross section from its minimum to the value at $s = (53\,\text{GeV})^2$ is almost 3 mb. The year after this phenomenological result was obtained, ISR data at this value of s became available, and it was found that the data gave an increase that is larger by about 50%. In view of the crudeness of the assumption (8.11), the result of figure 8.1 was considered to be a theoretical triumph.

For model 1 the ISR result on the pp total cross section is used as an input. This makes it unnecessary to introduce dubious assumptions such as equation (8.11). However, there is still no way to determine the value of c'. For this reason two separate sets of phenomenological parameters were obtained at this second stage, one set with $c' = 0$ and the other with $c' = 1$.

With the assumed values of c', six pieces of experimental data are needed to determine the other six parameters ($\lambda, c, f, x_0, A(pp)$, and $A(p\bar{p})$). These six pieces were chosen as follows:

1. At the laboratory energy $E = 20$ GeV, the pp total cross section is 39.0 mb [19].

Table 8.1
Phenomenological parameters for model 1[a]

Parameter	First Set of Parameters	Second Set of Parameters
c	0.082925	0.20193
c'	0.	1.
λ	0.60071	0.59582
$x_0(pp)$	3.8750	3.7508
$f(pp)$	3.1499	9.4966
$A(pp)$	47.372	9.4468
$A(p\bar{p})$	101.81	63.772

a. c' is chosen to be 0 for the first set of parameters and 1 for the second set.

2. At $E = 60$ GeV, the pp total cross section is 38.4 mb [19].

3. At $E = 1485$ GeV, which corresponds to $s = (53\ \text{GeV})^2$, the pp total cross section is 43.2 mb [11, 12].

4. At $E = 60$ GeV, the dip in the pp elastic differential cross section occurs at $|t| = 1.8\ (\text{GeV}/c)^2$ [20].

5. At $E = 505$ GeV, the average slope between $|t| = 0.14$ and $0.24\ (\text{GeV}/c)^2$ for $\ln d\sigma/dt$ is $11.5\ (\text{GeV}/c)^{-2}$ for pp elastic scattering [13].

6. At $E = 47.5$ GeV, the $p\bar{p}$ total cross section is 44.1 mb [19].

Because at that time the pp data were much better than $p\bar{p}$ data, five of the six pieces of data came from pp scattering. Of course, one piece of data from $p\bar{p}$ scattering is essential in order to determine the value of $A(p\bar{p})$.

The values of the parameters obtained [13, 14] at this second stage are tabulated in table 8.1. The fact that the parameters are given to five places does not mean such impossible accuracy but rather represents an attempt to reduce excessive roundoff error.

The determination of the parameters for model 2 is more complicated. The main idea is to carry out an overall fit using the available pp and $p\bar{p}$ data for the total cross section σ_{tot}, the ratio of the real to imaginary parts of the forward amplitude $\rho = \text{Re}\,\mathscr{M}(s, 0)/\text{Im}\,\mathscr{M}(s, 0)$, and the differential cross sections $d\sigma/dt$ mainly from the CERN ISR. However, in order to carry out this fitting, it is necessary to include Regge backgrounds. This is accomplished by using

$$\Omega(s, x_\perp^2) = S(s)F(x_\perp^2) + R(s, x_\perp^2), \tag{8.12}$$

instead of equation (8.5), where R is the Regge term. When the parametrization for the proton electromagnetic form factor is taken as

$$G(t) = \frac{1}{(1 - t/m_1^2)(1 - t/m_2^2)},$$ (8.13)

the values of the six parameters are found to be [3]

$c = 0.167,$

$c' = 0.748,$

$a = 1.953 \text{ GeV},$

$f = 7.115 \text{ GeV},$

$m_1 = 0.586 \text{ GeV},$

$m_2 = 1.704 \text{ GeV}.$ (8.14)

Note that unlike model 1 the value of c' is determined.

8.4 Results for Model 1

Phenomenological model 1 is completely described by equations (8.3) through (8.6), (8.7a), and (8.8) through (8.10) with the parameters given in table 8.1. It is so simple that numerical computations can be, and indeed were, carried out on desk calculators. Extensive numerical results, in the form of curves, were given over a decade ago [14]. Instead of reproducing these easily available curves, we concentrate on a few points.

Figure 8.2 [13] shows a comparison of the theoretical and experimental total cross sections at the time when the first set of parameters of table 8.1 was obtained. In addition to the pp and $p\bar{p}$ cross sections, figure 8.2 also shows the $\pi^{\pm}p$ and $K^{\pm}p$ total cross sections. The theoretical predictions for these $\pi^{\pm}p$ and $K^{\pm}p$ cross sections are somewhat less accurate than those for pp and $p\bar{p}$, primarily because of the inaccuracy of the experimental data, used as inputs, for laboratory energies at and below 60 GeV. A striking and unexpected prediction, later experimentally verified, is that, for momentum transfers of about 1 GeV/c and higher, the $K^{\pm}p$ elastic differential cross sections are much larger than those for pp, $p\bar{p}$, and $\pi^{\pm}p$.

Let us return to the pp and $p\bar{p}$ cross sections. Figure 8.3 shows the

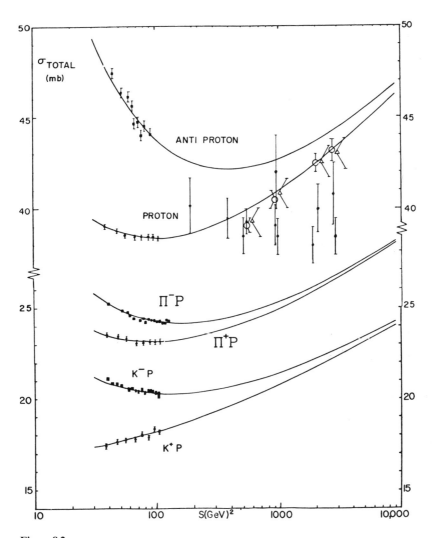

Figure 8.2
Comparison of the first fit with experimental data from Serpukhov [19], Fermilab [21], and
ISR [11, 12, 22].

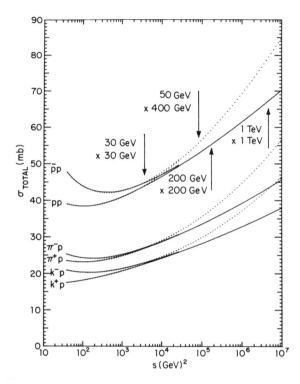

Figure 8.3
Comparison of the predictions of rising total cross section with the first UA4 data [4] from the CERN $p\bar{p}$ Collider. For $s > 10^4$ (GeV)^2 the theoretical pp and $p\bar{p}$ total cross sections are virtually the same, and the solid and dashed lines correspond, respectively, to the first and the second sets of parameters of table 8.1.

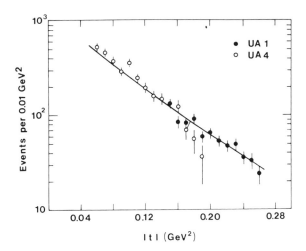

Figure 8.4
$|t|$-distribution of elastic scattering events in the $|t|$ range from 0.05 to 0.26 $(\text{GeV}/c)^2$ taken from [5] and compared to the $c' = 1$ prediction of model 1 [14].

theoretical total cross sections over a larger energy range, together with the first measurement [4] nine years later by the UA4 Collaboration, at the CERN $p\bar{p}$ Collider.

Finally, figure 8.4 shows a comparison of the unnormalized differential cross sections with the data from the UA1 [5] and UA4 [4] collaborations at the CERN $p\bar{p}$ Collider. The agreement is nearly perfect.

8.5   Results for Model 2

Because model 2 is at present the most accurate one for high energies, we give in some detail its results and predictions. As already mentioned, it would be interesting to compare these predictions with future data from the SSC and the CERN LHC.

Figure 8.5 shows the results of the ratio of the real to imaginary parts of the pp and $p\bar{p}$ forward scattering amplitudes. It is seen that, for both cases, this ratio is negative at low energies, increases to positive values, and eventually decreases gradually to zero. This qualitative behavior is a direct consequence of the increasing total cross section. Historically, when the

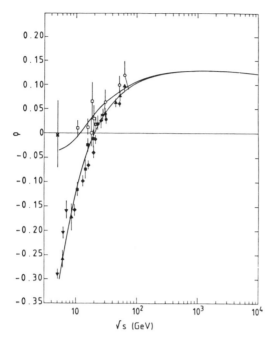

Figure 8.5
$\rho = \mathrm{Re}\,\mathcal{M}(s,0)/\mathrm{Im}\,\mathcal{M}(s,0)$ for pp (solid) and $p\bar{p}$ (open) (data from [23]) as a function of $\sqrt{s}$.

data from the Alternating Gradient Synchrotron at the Brookhaven National Laboratory first became available, there was a puzzle why this ratio seemed to approach zero too fast. The answer is, of course, the swing into positive values. The data shown are from various references [16, 23–28].

Figure 8.6 shows a comparison of the results of model 2 with data from the CERN ISR [25, 29, 30] (figure 8.6a) and from the $p\bar{p}$ Collider [4] (figure 8.6b). In both figures 8.5 and 8.6a, the Regge backgrounds have been included. At the high energy of $\sqrt{s} = 546$ GeV for figure 8.6b, the Regge backgrounds are negligible. The agreement is quite satisfactory; in particular, a shoulder is present at $|t| = 0.8$ $(\mathrm{GeV}/c)^2$ instead of a dip of the type seen in figure 8.6a.

Turning our attention to higher energies, we summarize in figure 8.7 the predictions of model 2 for σ_{tot}, $\sigma_{\mathrm{el}}/\sigma_{\mathrm{tot}}$, and ρ, together with the only available data points from the CERN $p\bar{p}$ Collider [4]. The ratio $\sigma_{\mathrm{el}}/\sigma_{\mathrm{tot}}$ is expected to reach 0.3 at $\sqrt{s} \sim 40$ TeV, and thereafter increases slowly to

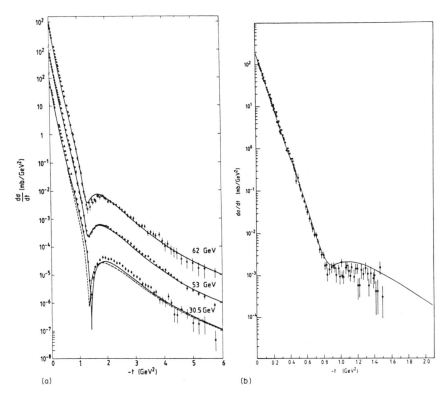

Figure 8.6
$d\sigma/dt$ for (a) pp at $\sqrt{s} = 30.5$, 53, and 62 GeV (data from [25, 29, 30]) and (b) $p\bar{p}$ at 546 GeV (data from [4]), compared with model 2. The prediction for $p\bar{p}$ at 30.5 GeV is the dashed curve in (a).

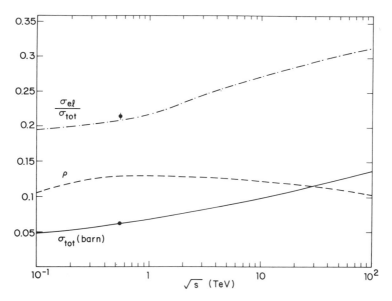

Figure 8.7
The total cross section σ_{tot} in barns, the ratio σ_{el}/σ_{tot} of the integrated elastic cross section to the total cross section, and the ratio ρ of the real to imaginary parts of the forward scattering amplitude for pp and $p\bar{p}$ scattering as functions of the center-of-mass energy $\sqrt{s}$. Note that both σ_{tot} and σ_{el}/σ_{tot} increase significantly in this energy range of 0.1 to 100 TeV, whereas ρ decreases slightly. The data points are from [4] by the UA4 Collaboration.

the asymptotic value of 0.5 [1]. In figure 8.7 and figure 8.8 (which gives the elastic differential cross section) the Regge backgrounds are negligible and hence not included. In other words, for such high energies, model 2 is described by equations (8.3) through (8.6), (8.7b), (8.8), and (8.9). More details of the differential cross sections for pp and $p\bar{p}$ elastic scattering are given in table 8.2. It can be seen from figure 8.8 and table 8.2 that the increases in the differential cross sections are more drastic for the higher energies. The first dip moves to smaller values [1] of $|t|$ as the energy increases, leading to a more pronounced structure because of the smallness of ρ, as shown in figure 8.7. In the meantime, the second dip structure is gradually restored. Another consequence of the movement of the dip is that the ratio of the differential cross section $d\sigma/dt$ in the nearly forward direction to that at the first shoulder between the first and second dips decreases with increasing energy. At 40 TeV this ratio is predicted to be only three-and-a-half orders of magnitude, compared with the five orders

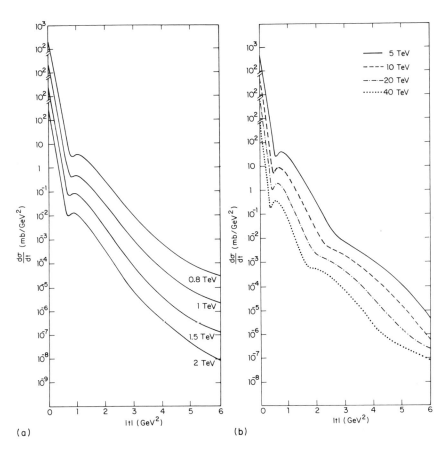

Figure 8.8
The pp and $p\bar{p}$ elastic differential cross section as a function of $|t|$, the square of the momentum transfer for energies in the ranges of (a) the CERN $p\bar{p}$ and the Fermilab $p\bar{p}$ colliders and (b) the CERN LHC and the SSC.

Table 8.2
pp and $p\bar{p}$ differential cross sections as given by model 2

$	t	(\text{GeV})^2$	\multicolumn{9}{l}{$d\sigma/dt$ [mb/(GeV)²] for various values of $\sqrt{s}$ (TeV)}	
	0.55	0.8	1	1.5	2	5	10	20	40	
0.05	8.70	9.70	1.04	1.16	1.26	1.63	1.94	2.29	2.67	× E + 2
0.10	4.05	4.45	4.71	5.20	5.57	6.81	7.76	8.66	9.45	× E + 1
0.15	1.96	2.12	2.22	2.41	2.54	2.92	3.14	3.26	3.26	
0.20	9.77	1.04	1.08	1.14	1.18	1.26	1.26	1.19	1.05	
0.25	4.98	5.19	5.30	5.45	5.51	5.38	4.89	4.06	2.99	× E + 0
0.30	2.57	2.62	2.63	2.62	2.58	2.25	1.80	1.24	7.03	× E − 1
0.35	1.34	1.33	1.31	1.25	1.19	9.00	6.08	3.33	1.86	
0.40	7.01	6.73	6.49	5.92	5.41	3.39	1.88	1.11	1.89	
0.45	3.66	3.38	3.17	2.73	2.37	1.20	6.89	1.06	2.85	
0.50	1.89	1.67	1.52	1.22	9.96	4.57	5.45	1.47	3.51	
0.55	9.63	8.05	7.05	5.23	4.07	2.79	6.82	1.78	3.69	
0.60	4.81	3.78	3.19	2.24	1.79	2.94	8.25	1.89	3.49	
0.65	2.35	1.74	1.43	1.08	1.08	3.48	8.98	1.84	3.09	
0.70	1.13	8.20	7.09	7.25	0.98	3.89	8.99	1.67	2.60	
0.75	5.49	4.43	4.56	6.90	1.07	4.04	8.46	1.46	2.12	
0.80	2.97	3.19	4.05	7.50	1.18	3.95	7.62	1.23	1.67	
0.85	2.01	3.00	4.24	8.11	1.24	3.69	6.64	1.01	1.29	
0.90	1.78	3.16	4.55	8.42	1.24	3.33	5.65	8.10	9.80	× E − 2
0.95	1.82	3.36	4.75	8.39	1.20	2.93	4.71	6.40	7.30	
1.00	1.92	3.47	4.78	8.06	1.12	2.53	3.86	4.99	5.37	
1.10	2.01	3.35	4.41	6.87	9.05	1.80	2.51	2.93	2.78	
1.20	1.89	2.93	3.71	5.44	6.90	1.23	1.57	1.65	1.37	
1.30	1.64	2.40	2.95	4.11	5.05	8.11	9.52	9.03	6.56	× E − 3
1.40	1.34	1.88	2.25	3.01	3.59	5.24	5.65	4.79	3.11	
1.50	1.06	1.43	1.67	2.15	2.50	3.33	3.28	2.49	1.58	
1.60	8.14	1.06	1.22	1.52	1.72	2.09	1.88	1.29	9.53	× E − 4
1.70	6.15	7.78	8.78	1.06	1.17	1.29	1.06	6.89	7.23	
1.80	4.59	5.64	6.26	7.30	7.86	7.93	5.95	4.00	6.40	
1.90	3.39	4.06	4.44	5.01	5.27	4.83	3.36	2.67	5.95	
2.00	2.49	2.91	3.13	3.43	3.52	2.93	1.95	2.06	5.50	
2.10	1.83	2.08	2.20	2.34	2.34	1.77	1.19	1.76	4.94	
2.20	1.34	1.48	1.55	1.59	1.55	1.08	7.93	1.57	4.31	
2.30	9.76	1.06	1.09	1.08	1.03	6.64	5.79	1.42	3.65	
2.40	7.14	7.56	7.64	7.37	6.84	4.19	4.61	1.26	3.01	
2.50	5.23	5.41	5.38	5.03	4.55	2.74	3.88	1.10	2.43	
2.60	3.84	3.88	3.80	3.44	3.04	1.89	3.36	9.42	1.93	
2.70	2.82	2.79	2.69	2.36	2.04	1.37	2.93	7.92	1.50	

Table 8.2 (continued)

|t|(GeV)²	$d\sigma/dt$ [mb/(GeV)²] for various values of $\sqrt{s}$ (TeV)									
	0.55	0.8	1	1.5	2	5	10	20	40	
2.80	2.09	2.01	1.91	1.63	1.38	1.04	2.56	6.56	1.16	× E − 4
2.90	1.55	1.46	1.36	1.13	9.47	8.33	2.21	5.35	8.79	× E − 5
3.00	1.15	1.06	9.80	7.92	6.56	6.86	1.89	4.31	6.60	
3.20	6.52	5.75	5.16	4.00	3.31	4.87	1.34	2.70	3.62	
3.40	3.77	3.20	2.80	2.12	1.79	3.50	9.08	1.64	1.93	
3.60	2.25	1.84	1.58	1.18	1.04	2.48	5.95	9.61	1.01	
3.80	1.38	1.09	9.26	6.93	6.44	1.72	3.79	5.51	5.23	× E − 6
4.00	8.75	6.75	5.65	4.26	4.14	1.16	2.35	3.09	2.77	
4.20	5.73	4.32	3.57	2.71	2.72	7.59	1.43	1.71	1.54	
4.40	3.87	2.86	2.34	1.77	1.80	4.86	8.47	9.35	9.23	× E − 7
4.60	2.69	1.96	1.59	1.18	1.19	3.04	4.93	5.11	6.08	
4.80	1.93	1.38	1.11	8.04	7.87	1.86	2.81	2.83	4.33	
5.00	1.42	1.00	7.97	5.55	5.19	1.10	1.57	1.62	3.25	
5.20	1.08	7.52	5.90	3.90	3.43	6.33	8.53	9.71	2.51	
5.40	8.34	5.78	4.48	2.82	2.29	3.49	4.54	6.25	1.94	
5.60	6.60	4.56	3.51	2.10	1.56	1.83	2.36	4.33	1.50	
5.80	5.33	3.67	2.82	1.62	1.10	0.90	1.21	3.19	1.14	
6.00	4.38	3.02	2.32	1.30	0.82	0.39	0.63	2.46	8.61	× E − 8

of magnitude at the CERN $p\bar{p}$ Collider [5]. A number of other features are also evident.

8.6 Discussion

The theoretical developments of the preceding chapters involve little assumption beyond the validity of gauge theories and the summation of perturbation series. By contrast, the choices of the models in this chapter depend heavily on the assumption of simplicity. In particular, this assumption has been used at least three times: in writing equations (8.5), (8.6), and (8.7). Although equation (8.6) is chosen for its simplicity, there does not seem to be another natural choice that is substantially different. On the other hand, the choice of the factorized form (8.5) is without doubt a less justified assumption; it is used because of our inability to find a natural and manageable alternative. We hope that in the future more elaborate and accurate models will be developed in which this assumption of factorization, equation (8.5), is either confirmed or replaced by a more refined form.

The models 0, 1, and 2 described here are not the only models with rising total cross sections. We describe briefly several others that deal with total and elastic cross sections. Even earlier than model 0, there was an attempt [31] to incorporate our result [1] from field theory. In this work the available experimental data on total cross sections were fitted under the constraint of the asymptotic behavior in the form $\{\ln[s^c/(\ln s)^{c'}]\}^2$. This leads to the striking feature that the $\pi^\pm p$ total cross sections increase rapidly for energies just beyond that available at Serpukhov [$\sqrt{s} \sim 10$ GeV], whereas those for pp and $p\bar{p}$ increase only slowly. Consequently, at high energies the $\pi^\pm p$ total cross section is much larger than those of pp and $p\bar{p}$. This approach did not yield any phenomenological information about differential cross sections.

Another approach is based on the droplet model [32]. Originally, it was a cornerstone of the droplet model that total cross sections approach constants at infinite energy. When it became clear that this was not so [11–14], efforts were made to allow the total cross sections to be slowly varying functions of energy. In all these efforts [33] the total cross sections are inputs; that is, various quantities are obtained as functions of the total cross sections and not directly as functions of the energy. Because the resulting elastic amplitudes are purely imaginary, the differential cross sections obtained in this way have zeros [33] instead of dips, as shown, for example, in figures 8.6 and 8.8.

There are many other phenomenological analyses of the rising total cross sections, ranging from pure data fitting to $SU(3)$. We give some examples of such efforts [34]. The relation of the results from quantum field theory to multiperipheral models has also been studied [35].

In this chapter we have concentrated primarily on one aspect of the comparison with experimental data, namely, on pp and $p\bar{p}$ elastic scattering. This choice is influenced by the availability of the data from the CERN $p\bar{p}$ Collider, nearly a decade after the phenomenological prediction. Actually, as discussed at the Tbilisi Conference [36], many comparisons are possible and desirable. Other examples include the violation of KNO scaling [37, 38], proton-nucleus scattering [39], and charm production [40]. However, prospects of reaching the multi-TeV energy ranges are good only for $p\bar{p}$ (and possibly pp) colliding beam accelerators. One of the most interesting future comparisons is perhaps $p\bar{p}$ diffractive dissociation at CERN, Fermilab, and SSC [14, 41].

If this multi-TeV region is referred to as the asymptopia, at least so far as pp and $p\bar{p}$ elastic scattering is concerned, then asymptopia is full of interesting structures and physics. Both theoretically and experimentally there are also many issues that deserve to be clarified. Perhaps the most obvious one concerns the possible characterization of the rapidly increasing inelastic cross section. Presumably both diffractive and pionization processes are involved, and a more refined analysis can be expected to lead to a deeper understanding.

References*

[1] H. Cheng and T. T. Wu, *Physical Review Letters* 24 (1970):1456.

[2] C. Bourrely, J. Soffer, and T. T. Wu, *Physics Letters B* 121 (1983):284.

[3] C. Bourrely, J. Soffer, and T. T. Wu, *Nuclear Physics B* 247 (1984):15, *Physical Review Letters* 54 (1985):757; CERN preprint TH-4643 (1987).

[4] UA4 Collaboration (R. Battiston, M. Bozzo, P. L. Braccini, et al.), *Physics Letters B* 115 (1982):333, 117 (1982):126, 127 (1983):472; CERN preprints EP84-90 and EP84-91 (1984).

[5] UA1 Collaboration (G. Arnison, A. Astbury, B. Aubert, et al.), *Physics Letters B* 121 (1983):77.

[6] G. Hanson, G. S. Abrams, A. M. Boyarski, et al., *Physical Review Letters* 35 (1975):1609.

[7] S. L. Wu, *Physics Reports* 107 (1984):59. See also reference [12] of chapter 1.

[8] TASSO Collaboration (R. Brandelik, W. Braunschweig, K. Gather, et al.), *Physics Letters B* 97 (1980):453; PLUTO Collaboration (C. Berger, H. Genzel, R. Grigull, et al.), *Physics Letters B* 97 (1980):459; CELLO Collaboration (H.-J. Behrend, C. Chen, J. H. Field, et al.), *Physics Letters B* 110 (1982):329.

[9] G. 't Hooft, paper presented at the Marseille meeting, 1972 (unpublished); D. J. Gross and F. Wilczek, *Physical Review Letters* 30 (1973):1343; H. D. Politzer, *Physical Review Letters* 30 (1973):1346.

[10] H. Cheng, J. K. Walker, and T. T. Wu, Paper 524, 16th International Conference on High Energy Physics, 1972 (unpublished).

[11] S. R. Amendolia, G. Bellettini, P. L. Braccini, et al., *Physics Letters B* 44 (1973):119.

[12] U. Amaldi, R. Biancastelli, C. Bosio, et al., *Physics Letters B* 44 (1973):112.

[13] H. Cheng, J. K. Walker, and T. T. Wu, *Physics Letters B* 44 (1973):97.

[14] T. T. Wu and H. Cheng, in *High Energy Collisions*, C. Quigg, ed. (New York: American Institute of Physics, 1973).

[15] H. Cheng, J. K. Walker, and T. T. Wu, *Physics Letters B* 44 (1973):283.

[16] V. Bartenev, R. A. Carrigan Jr., I.-H. Chiang, et al., *Physical Review Letters* 31 (1973): 1367.

[17] T. Regge, *Nuovo Cimento* 14 (1959):951, 18 (1960):947.

[18] C. Bourrely, J. Soffer, and T. T. Wu, *Physical Review D* 19 (1979):3249.

[19] S. P. Denisov, S. V. Donskov, Yu. P. Gorin, et al., *Physics Letters B* 36 (1971):415.

[20] J. V. Allaby, A. N. Diddens, R. W. Dobinson, et al., *Physics Letters B* 34 (1971):431.

*Complete references are given in the back of this book.

[21] G. Charlton, Y. Cho, M. Derrick, et al., *Physical Review Letters* 29 (1972): 515; F. T. Dao, D. Gordon, J. Lach, et al., *Physical Review Letters* 29 (1972): 1627; J. W. Chapman, N. Green, B. P. Roe, et al., *Physical Review Letters* 29 (1972): 1686.

[22] Aachen-CERN-Harvard-Geneva-Torino Collaboration, paper presented by C. Rubbia at the 16th International Conference on High Energy Physics, 1972 (unpublished).

[23] K. J. Foley, R. S. Jones, S. J. Lindenbaum, et al., *Physical Review Letters* 19 (1967): 857.

[24] G. G. Beznogikh, A. Bujak, L. F. Kirillova, et al., *Physics Letters B* 39 (1972): 411.

[25] U. Amaldi, G. Cocconi, A. N. Diddens, et al., *Physics Letters B* 66 (1977): 390.

[26] L. A. Fajardo, R. Majka, J. N. Marx, et al., *Physical Review D* 24 (1981): 46.

[27] M. Ambrosio, G. Anzivino, G. Barbarino, et al., *Physics Letters B* 115 (1982): 495.

[28] N. Amos, M. M. Block, G. J. Bobbink, et al., *Physics Letters B* 120 (1983): 460.

[29] G. Barbiellini, M. Bozzo, P. Darriulat, et al., *Physics Letters B* 39 (1972): 663; A. Böhm, M. Bozzo, R. Ellis, et al., *Physics Letters B* 49 (1974): 491; L. Baksay, L. Baum, A. Böhm, et al., *Nuclear Physics B* 148 (1978): 1.

[30] N. Kwak, E. Lohrmann, E. Nagy, et al., *Physics Letters B* 58 (1975): 233, 62 (1976): 363, 68 (1977): 374, *Nuclear Physics B* 150 (1979): 221.

[31] V. Barger and R. J. N. Phillips, *Nuclear Physics B* 32 (1971): 93, 40 (1972): 205.

[32] T. T. Wu and C. N. Yang, *Physical Review* 137 (1965): B708; N. Byers and C. N. Yang, *Physical Review* 142 (1966): 976; T. T. Chou and C. N. Yang, in *High Energy Physics and Nuclear Structure* (Amsterdam: North-Holland, 1967), 348–359, *Physical Review Letters* 20 (1968): 1213, *Physical Review* 170 (1968): 1591, 175 (1968): 1832, *Physical Review Letters* 25 (1970): 1072; J. Benecke, T. T. Chou, C. N. Yang, and E. Yen, *Physical Review* 188 (1969): 2159; J. Benecke, *Nuovo Cimento A* 2 (1971): 615; J. P. Hsu, *Physical Review D* 3 (1971): 257.

[33] A. W. Chao and C. N. Yang, *Physical Review D* 8 (1973): 2063; D. J. Clarke and S. Y. Lo, *Physical Review D* 10 (1974): 1519, *Physics Letters B* 87 (1979): 379; F. Hayot and U. P. Sukhatme, *Physical Review D* 10 (1974): 2183; R. Carlitz, *Physical Review D* 10 (1974): 3878; T. T. Chou and C. N. Yang, *Physical Review D* 19 (1979): 3268, *Physical Review Letters* 46 (1981): 764.

[34] R. E. Hendrick, P. Langacker, B. E. Lautrup, S. J. Orfanidis, and V. Rittenberg, *Physical Review D* 11 (1975): 536; E. Rabinovici, *Physical Review D* 11 (1975): 1969; C. B. Chiu and K. H. Wang, *Nuovo Cimento A* 27 (1975): 213; C. Quigg and E. Rabinovici, *Physical Review D* 13 (1976): 2525; S. Y. Chu, B. R. Desai, B. C. Shen, and R. D. Field, *Physical Review D* 13 (1976): 2967; P. E. Volkovitskii and A. M. Lapidus, *Soviet Journal of Nuclear Physics* 31 (1980): 250, 380.

[35] M. Wakano, *Progress of Theoretical Physics* 53 (1975): 761.

[36] T. T. Wu, *Proceedings of the 18th International Conference on High Energy Physics*, 1976, vol. 1, A5–A23.

[37] H. M. Fried and C. I. Tan, *Physical Review D* 9 (1974): 314.

[38] Z. Koba, H. B. Nielsen, and P. Olesen, *Nuclear Physics B* 40 (1972): 317.

[39] P. M. Fishbane, J. G. Schaffner, and J. S. Trefil, *Physical Review D* 10 (1974): 3056; J. Pumplin, *Physical Review D* 10 (1974): 3736.

[40] L. M. Jones and H. M. Wyld, *Physical Review D* 17 (1978): 1782.

[41] H. Cheng, J. K. Walker, and T. T. Wu, *Physical Review D* 9 (1974): 749; P. K. Williams, *Physical Review D* 17 (1978): 909.

9 Scattering of Particles with Structure

9.1 Introduction

On the basis of gauge field theories, we have studied high-energy hadronic collision processes with large cross sections. For the purpose of elucidating the physical contents of these results, the first eight chapters are devoted primarily to the main issues and the essential techniques. This necessarily leaves many questions unanswered. In the remainder of this book we attempt to discuss some of these problems and also give more technical details.

The incident particles so far have been assumed to be fermions with no internal structure. Because hadrons clearly have interesting internal structures, as evidenced, for example, by their form factors, we discuss in this chapter the role of such structures. This problem is most easily approached by considering scattering processes in which one or both of the incident particles are vector mesons.

As we will find, all the qualitative features appearing in the high-energy scattering of fermions also appear in the scattering of vector mesons. But the scattering of vector mesons has one additional feature. In an Abelian gauge field theory, a vector meson cannot emit another vector meson. In order to exchange vector mesons with the other incident particle, a vector meson must first turn into one or more fermion-antifermion pairs. (We sometimes abbreviate the vector meson, the fermion, and the antifermion by V, f, and $\bar{f}$, respectively.) The scattering of a vector meson can thus be viewed as the scattering of a composite particle, consisting of a distribution of fermions and antifermions, which exchanges vector mesons with the other incident particle. As a result, the scattering of vector mesons provides a good model for the scattering of hadrons, which are conceived of as bound states of quarks and gluons.

For the benefit of those who are not interested in the details of calculations, we first summarize the results [1]. Let us recall that in chapter 3 we found that the lowest-order amplitude for fermion-fermion scattering is real and consists of one-vector-meson exchange and that the lowest-order amplitude that has an imaginary part is the two-vector-meson exchange amplitude given by expression (3.22). For fermion–vector meson elastic scattering the exchange of one vector meson is forbidden by Furry's theorem [2]. Thus the lowest-order amplitude is that for two-vector-meson exchange and it is purely imaginary. Furthermore, the amplitude for two-vector-meson exchange takes a form that is a generalization of expression

(3.22). More specifically, we can write, for the two-vector-meson exchange amplitude of $a + b \to a + b$,

$$
\mathcal{M} \sim i2(p_a \cdot p_b) \int \frac{d^2 q_\perp}{(2\pi)^2} \frac{\mathcal{I}^a(\Delta, \mathbf{q}_\perp) \cdot \mathcal{I}^b(\Delta, \mathbf{q}_\perp)}{(\mathbf{q}_\perp^2 + \lambda^2)[(\Delta - \mathbf{q}_\perp)^2 + \lambda^2]}, \tag{9.1}
$$

where p_a and p_b are the incident momenta of particle a and partible b, respectively, and $\mathcal{I}^a$ (resp. $\mathcal{I}^b$) is called the impact factor for particle a (resp. b). Specifically, for a fermion or an antifermion, the impact factor is

$$
\mathcal{I}^f = \frac{g^2}{2m}, \tag{9.2}
$$

whereas for a vector meson

$$
\mathcal{I}^V(\Delta, \mathbf{q}_\perp) = \frac{g^4}{4\pi^3} \int d^2 p_\perp \int_0^1 d\beta
$$

$$
\times \left\{ \frac{\beta(1 - \beta)(\beta^2 \Delta_i \Delta_j - 4\mathbf{p}_{\perp i}\mathbf{p}_{\perp j}) - [\tfrac{1}{2}\beta^2 \Delta^2 - \beta(1 - \beta)\lambda^2]\delta_{ij}}{[(\mathbf{p}_\perp + \tfrac{1}{2}\beta\Delta)^2 + m^2 - \beta(1 - \beta)\lambda^2][(\mathbf{p}_\perp - \tfrac{1}{2}\beta\Delta)^2 + m^2 - \beta(1 - \beta)\lambda^2]} \right.
$$

$$
\left. - \frac{4\beta(1 - \beta)(Q_i Q_j - \mathbf{p}_{\perp i}\mathbf{p}_{\perp j}) - [2Q^2 - \beta(1 - \beta)\lambda^2]\delta_{ij}}{[(\mathbf{p}_\perp + Q)^2 + m^2 - \beta(1 - \beta)\lambda^2][(\mathbf{p}_\perp - Q)^2 + m^2 - \beta(1 - \beta)\lambda^2]} \right\}, \tag{9.3}
$$

where

$$
Q = \tfrac{1}{2}\mathbf{q}_\perp - \tfrac{1}{2}\beta\Delta. \tag{9.4}
$$

The integral in equation (9.3) can be cast into a different form. By introducing a Feynman parameter x and carrying out the integration over $\mathbf{p}_\perp$, we obtain after some algebra

$$
\mathcal{I}^V(\Delta, \mathbf{q}_\perp) = \frac{g^4}{2\pi^2} \int_0^1 dx \int_0^1 d\beta
$$

$$
\times \left[\frac{2\beta^3(1 - \beta)x(1 - x)\Delta_i\Delta_j - \tfrac{1}{4}\beta^2\Delta^2[1 - 8\beta(1 - \beta)(x - \tfrac{1}{2})^2]\delta_{ij} + \beta(1 - \beta)\lambda^2\delta_{ij}}{\beta^2 x(1 - x)\Delta^2 + m^2 - \beta(1 - \beta)\lambda^2} \right.
$$

$$
\left. - \frac{8\beta(1 - \beta)x(1 - x)Q_i Q_j - Q^2[1 - 8\beta(1 - \beta)(x - \tfrac{1}{2})^2]\delta_{ij} + \beta(1 - \beta)\lambda^2\delta_{ij}}{4x(1 - x)Q^2 + m^2 - \beta(1 - \beta)\lambda^2} \right]. \tag{9.5}
$$

We call relation (9.1) the impact factor representation. If we replace $\mathcal{I}^a$ and $\mathcal{I}^b$ in relation (9.1) by $\mathcal{I}^f$, we obtain expression (3.22), the two-vector-meson exchange amplitude for fermion-fermion scattering. If we replace $\mathscr{I}^a$ by $\mathscr{I}^V$ and $\mathscr{I}^b$ by $\mathscr{I}^f$, we obtain the two-vector-meson exchange amplitude for vector meson–fermion scattering, which is calculated in the next section. We can similarly obtain the two-vector-meson exchange amplitude for vector meson–vector meson scattering, vector meson–antifermion scattering, fermion-antifermion scattering, and antifermion-antifermion scattering.

The impact factor $\mathscr{I}^f$ is simply a constant. This is because the fermion is treated as a point particle. On the other hand, $\mathscr{I}^V$ is a function of Δ and $\mathbf{q}_\perp$. This is because a vector meson behaves like a composite particle, as mentioned before.

Next let us recall that, for fermion-fermion scattering, not only the two-vector-meson exchange amplitude but also the sum of multi-vector-meson exchange amplitudes are known. The latter is simply given by equation (3.36). A generalization of equation (3.36) to fermion–vector meson scattering is also straightforward. Let us define

$$P_\pm(\mathbf{q}_\perp) \equiv \pm \frac{1}{iy^2} \int d^2x_\perp e^{\mathbf{q}_\perp \cdot \mathbf{x}_\perp} \left\{ \exp\left[\pm \frac{ig^2}{2\pi} K_0(\lambda|\mathbf{x}_\perp|) \right] - 1 \right\}. \tag{9.6}$$

[Note that the lowest-order term in identity (9.6) is simply $(\mathbf{q}_\perp^2 + \lambda^2)^{-1}$.] Then the fermion-fermion multi-vector-meson exchange amplitude (3.36) is

$$-\frac{s}{2m^2} g^2 P_-(\Delta).$$

As we show later, the fermion–vector meson multi-vector-meson exchange amplitude is given by

$$is \int \frac{d^2q_\perp}{(2\pi)^2} \mathscr{I}^V(\Delta, \mathbf{q}_\perp) \mathscr{I}^f P_-(\mathbf{q}_\perp) P_+(\Delta - \mathbf{q}_\perp). \tag{9.7}$$

Each of the P-functions in expression (9.7) describes the interaction of a particle in the created fermion-antifermion pair with the incident fermion. An alternative form for expression (9.7) is

$$is \int \frac{d^2q_\perp}{(2\pi)^2} \mathscr{I}^V(\Delta, \mathbf{q}_\perp) \mathscr{I}^f [\mathscr{S}_-(\mathbf{q}_\perp)\mathscr{S}_+(\Delta - \mathbf{q}_\perp) - g^{-4}(2\pi)^4 \delta^{(2)}(\mathbf{q}_\perp) \delta^{(2)}(\Delta - \mathbf{q}_\perp)], \tag{9.8}$$

where

$$\mathscr{S}_\pm(\mathbf{q}_\perp) \equiv P_\pm(\mathbf{q}_\perp) \pm (ig^2)^{-1}(2\pi)^2 \delta^{(2)}(\mathbf{q}_\perp). \tag{9.9}$$

We can derive expression (9.8) from expression (9.7) by utilizing the fact that

$$\mathscr{I}^V(\Delta, 0) = \mathscr{I}^V(\Delta, \Delta) = 0, \tag{9.10}$$

which is easily proved from equation (9.3). Another property of $\mathscr{I}^V$, also easily derived from equation (9.3), is

$$\mathscr{I}^V(\Delta, \mathbf{q}_\perp) = \mathscr{I}^V(\Delta, \Delta - \mathbf{q}_\perp). \tag{9.11}$$

It is also possible to express the V-V multi-vector-meson exchange amplitude in a representation similar to expression (9.8). For V-V scattering, each of the incident vector mesons turns into a fermion-antifermion pair, which exchanges vector mesons with the other fermion-antifermion pair. The scattering amplitude is

$$\frac{1}{2} is \int \prod_{n=1}^{4} \frac{d^2 q_{n\perp}}{(2\pi)^4} (2\pi)^2 \delta^{(2)} \left(\sum_1^4 \mathbf{q}_{n\perp} - \Delta \right) \mathscr{I}^V(\Delta, \mathbf{q}_{1\perp} + \mathbf{q}_{2\perp}) \mathscr{I}^V(\Delta, \mathbf{q}_{1\perp} + \mathbf{q}_{3\perp}) g^4$$

$$\times \left[\mathscr{S}_+(\mathbf{q}_{1\perp}) \mathscr{S}_-(\mathbf{q}_{2\perp}) \mathscr{S}_-(\mathbf{q}_{3\perp}) \mathscr{S}_+(\mathbf{q}_{4\perp}) - \prod_1^4 \left(\frac{2\pi}{g} \right)^2 \delta^{(2)}(\mathbf{q}_{n\perp}) \right]. \tag{9.12}$$

Each P-function (contained in $\mathscr{S}$) represents the interaction of a particle in a pair with a particle in another pair.

The impact-factor representations (9.8) and (9.12) can be regarded as generalizations of the Glauber form [3]. To see this, we transform the integrals into ones over transverse distances other than transverse momenta. Let

$$I^V(\Delta, \mathbf{x}_\perp) \equiv \frac{1}{g^2} \int \frac{d^2 q_\perp}{(2\pi)^2} \exp[-i(\mathbf{q}_\perp - \tfrac{1}{2}\Delta) \cdot \mathbf{x}_\perp] \mathscr{I}^V(\Delta, \mathbf{q}_\perp), \tag{9.13}$$

and

$$I^f \equiv 1/m. \tag{9.14}$$

Then expression (9.7) becomes

$$\frac{1}{2} is \int d^2 b\, d^2 x_\perp \exp(i\Delta \cdot \mathbf{b}) I^V(\Delta, \mathbf{x}_\perp) I^f$$

$$\times \left\{ 1 - \exp\left[\frac{ig^2}{2\pi} K_0(\lambda |\mathbf{b} + \tfrac{1}{2}\mathbf{x}_\perp|) - \frac{ig^2}{2\pi} K_0(\lambda |\mathbf{b} - \tfrac{1}{2}\mathbf{x}_\perp|) \right] \right\}. \tag{9.15}$$

Let us compare expression (9.15) with the Glauber form for the pion-deuteron elastic scattering amplitude. (The deuteron is considered to be a proton and a neutron bound together, and pion-deuteron scattering is considered to be the scattering of the pion with the two constituent nucleons.) This Glauber form is

$$\frac{ik}{2\pi}\int d^2b\,d^3x\exp(i\mathbf{\Delta}\cdot\mathbf{b})\phi_f^*(\mathbf{x})\phi_i(\mathbf{x})\{1-\exp[i\chi_n(\mathbf{b}-\tfrac{1}{2}\mathbf{x}_\perp)+i\chi_p(\mathbf{b}+\tfrac{1}{2}\mathbf{x}_\perp)]\},$$

$$(9.16)$$

where ϕ_i (resp. ϕ_f) is the wave function for the internal state of the incoming (resp. outgoing) deuteron, $k\mathbf{e}_3$ is the momentum of the incoming pion, and χ_n and χ_p are the phase shifts associated with the pion-neutron and pion-proton scattering, respectively, and are the counterparts of $(ie^2/2\pi)K_0$ and $(-ie^2/2\pi)K_0$ in expression (9.15). Let us define the deuteron impact factor as

$$I^D(\mathbf{x}_\perp)=\int_{-\infty}^{\infty}dz\,\phi_f^*(\mathbf{x}_\perp,z)\phi_i(\mathbf{x}_\perp,z),\qquad(9.17)$$

where, for clarity, we have written $\phi_i(\mathbf{x}_\perp,z)$ and $\phi_f(\mathbf{x}_\perp,z)$ in place of $\phi_i(\mathbf{x})$ and $\phi_f(\mathbf{x})$. Then expression (9.16) is seen to be in the form of the impact-factor representation (9.15) but with one difference: I^D as given by equation (9.17) is a function of $\mathbf{x}_\perp$ only, whereas I^V depends, in addition, on the momentum transfer. This difference is due to the neglect in the Glauber form of the relativistic effects of the recoil of the deuteron [4]. More precisely, equation (9.17) tells us that the deuteron impact factor is related to the overlapping of the internal wave functions of the deuteron. There are two modifications we must make in a fully relativistic theory:

1. The function $\phi_f(\mathbf{x})$ describes the internal state of the deuteron *in the Lorentz frame in which the outgoing deuteron is at rest*. When Δ is comparable to the deuteron mass, the wave function of the deuteron as it appears to an observer at rest in the laboratory system is not $\phi_f(\mathbf{x})$ but is related to $\phi_f(\mathbf{x})$ by a Lorentz transformation.

2. The wave functions in equation (9.17) should *not* be interpreted as the wave functions at $t=0$. Rather, they are wave functions on the light cone $t-z=0$ if the momentum of the pion is taken to be in the direction of the positive z-axis. (If the momentum of the pion is taken to be in the direction of the negative z-axis, then ϕ_i and ϕ_f in equation (9.17) are wave functions on $t+z=0$.) This is because the pion is traveling with a velocity near that of light and the deuteron is therefore hit at time $t\sim z$.

With these two observations, let us designate the wave function for the internal state of the outgoing deuteron as seen by an observer in the laboratory system to be $\psi_f^*(\Delta, \mathbf{x}_\perp, z)$; then, instead of equation (9.17), the deuteron impact factor should read

$$I^D(\Delta, \mathbf{x}_\perp) = \frac{1}{2} \int_{-\infty}^{\infty} dx_+ \, \psi_f^*(\Delta, \mathbf{x}_\perp, z)\phi_i(\mathbf{x}_\perp, z)$$

$$= \int_{-\infty}^{\infty} dz \, \psi_f^*(\Delta, \mathbf{x}_\perp, z)\phi_i(\mathbf{x}_\perp, z), \tag{9.18}$$

where the wave functions are understood to be those on the light cone $t - z = 0$. From equation (9.18), we see that the deuteron impact factor is dependent on Δ. When Δ is small compared with the deuteron mass, this relativistic effect of recoil is negligible. However, in hadron-hadron scattering with Δ equal to at least 1 GeV/c, the recoil effect must be taken into account. In principle, the wave functions of hadrons can be calculated from QCD, but such calculations are as yet beyond our means. An approximation has been given by Low [5] that makes use of the bag model [6].

The considerations given suggest that we can use expression (9.12) to describe vector meson–vector meson scattering. In the approximation of treating a vector meson as a $q\bar{q}$ pair (where q denotes a quark and $\bar{q}$ an antiquark), we must identify $\mathscr{S}_+$ and $\mathscr{S}_-$ in expression (9.12) as the $q\bar{q}$ scattering amplitude and the qq scattering amplitude, respectively. The distribution of q and $\bar{q}$ in a vector meson is described by an impact factor. The inclusion of states such as $q\bar{q}g$ (where g denotes a gluon) in the vector meson can be accomplished by using additional impact factors and the gq, $g\bar{q}$, and gg scattering amplitudes. Similarly, a proton is approximated by a bound state of three quarks, and a straight-forward extension of expression (9.12) holds for the amplitude involving the scattering of protons.

For completeness, let us discuss the applications of the impact-factor representation to QED. For this purpose we replace λ by 0 and interpret g^2 as e^2, the fermion as the electron, and the vector meson as the photon. The impact-factor representation then holds for $e\gamma$ and $\gamma\gamma$ scattering. It also holds for the scattering of an electron in an external field or for that of a photon in an external field (Delbrück scattering). In these cases we can think of the external field as a static nucleus with the impact factor

$$\mathscr{I}^N = \frac{Z^2 e^2}{2M}, \tag{9.19}$$

where M and Z are the mass and the charge of the nucleus, respectively. Also, if b is a static nucleus, its four-momentum has only a time component M. Thus s is replaced by

$$2p_a p_b \sim 2\omega M,$$

where ω is the energy of particle a. The Delbrück scattering amplitude [7], with the external field taken care of to *all* orders, takes the form

$$i\omega Z^2 e^2 \int \frac{d^2 q_\perp}{(2\pi)^2} \frac{[\mathscr{I}^V(\mathbf{\Delta}, \mathbf{q}_\perp)]_{\lambda=0}}{(\mathbf{q}_\perp^2)^{1+i\alpha Z}[(\mathbf{\Delta} - \mathbf{q}_\perp)^2]^{1-i\alpha Z}}. \tag{9.20}$$

Expression (9.20) is obtained from expression (9.7) by making the replacement given by equation (9.19), with $P_-(\mathbf{q}_\perp)P_+(\mathbf{\Delta} - \mathbf{q}_\perp)$ replaced by

$$(\mathbf{q}_\perp^2)^{-1-i\alpha Z}[(\mathbf{\Delta} - \mathbf{q}_\perp)^2]^{-1+i\alpha Z},$$

the last replacement being valid in the limit of $\lambda \to 0$ and with $g^2 = Ze^2$. Expression (9.20) is of experimental interest. It is significant that expression (9.20) holds even if αZ is of the order of 1. Thus it is valid even if the nucleus has a large Z value. This expression has been verified experimentally [8]. Other aspects of Delbrück scattering, both theoretical [9] and experimental [10], have also been studied.

The impact-factor representation (9.1) for $\gamma\gamma$ scattering yields one interesting result. By setting $\mathbf{\Delta} = 0$ in this amplitude and by making use of the optical theorem, viz.,

$$\sigma(s) = s^{-1}\mathscr{M}_{\gamma\gamma}(\mathbf{\Delta} = 0),$$

where $\sigma(s)$ is the total cross section for photon-photon scattering, we get, up to the eighth order of e,

$$\lim_{s \to \infty} \sigma(s) \sim \frac{\alpha^4}{36\pi m^2}[175\zeta(3) - 38] \sim 6.5 \ \mu\text{b}, \tag{9.21}$$

independent of s as well as the helicities of the incoming photons [11].

The cross section (9.21) is due to $\gamma\gamma \to e^+e^-e^+e^-$. The corresponding cross section resulting from $\gamma\gamma \to \mu^+\mu^-\mu^+\mu^-$ is, of course, given by relation (9.21) but with m reinterpreted as the μ mass. This cross section is thus smaller by five orders of magnitude. The contribution from $\gamma\gamma \to e^+e^-\mu^+\mu^-$ has been studied by Masujima [12].

9.2 $f + V \to f + V$: **Second and Fourth Orders**

In the remainder of this chapter we present in some detail the calculation of the high-energy scattering amplitude for fermion–vector meson scattering resulting from multi-vector-meson exchange. That for vector meson–vector meson scattering can be carried out in a similar manner [1] but is not given here.

The lower-order diagrams for fermion–vector meson scattering are the two diagrams illustrated in figure 9.1. Both diagrams give amplitudes of the order of s^0 in the high-energy limit and can therefore be neglected. To see this, consider figure 9.1a. We take the laboratory system in which the fermion is initially at rest. Then the scattering amplitude corresponding to this diagram is

$$g^2 \bar{u}(\Delta) \frac{\gamma_j (\not{k}_1 + \not{p}_1 + m) \gamma_i}{s - m^2} u(0). \tag{9.22}$$

In the limit $s \to \infty$, we have

$$k_1 \sim \left[\frac{s}{2m}, \frac{s}{2m}, 0, 0 \right]. \tag{9.23}$$

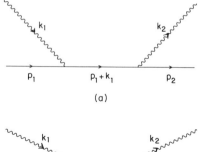

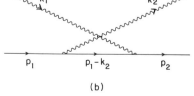

Figure 9.1
The lowest-order diagrams for elastic fermion–vector meson scattering.

Thus

$$\not{k}_1 \sim \frac{s}{2m}(\gamma_0 - \gamma_3).$$ (9.24)

Expression (9.22) therefore asymptotically approaches

$$\frac{g^2}{2m}\bar{u}(\Delta)\gamma_j(\gamma_0 - \gamma_3)\gamma_i u(0),$$ (9.25)

independent of s. Similar considerations apply to the diagram in figure 9.1b.

The fourth-order diagrams for fermion–vector meson scattering are illustrated in figure 9.2. The sum of the seven diagrams (figures 9.2a through 9.2g) gives an amplitude of the order of s^0 (logarithmic factors of s ignored). In fact, aside from the renormalization of mass, charge, and the wave functions, this amplitude is equal to expression (9.22) with the vertex

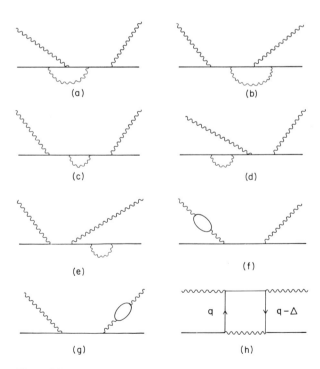

Figure 9.2
Fourth-order diagrams for elastic fermion–vector meson scattering.

functions and the fermion propagator modified by their lowest-order radiative corrections.

The eighth diagram (figure 9.2h) is slightly more complicated [13]. We
can view the process as one in which the fermions of momenta q and $q - \Delta$
are being exchanged between the incident particles. Because fermions are
of spin $\frac{1}{2}$, the rule (7.2) shows that the corresponding amplitude is of the
order of s^0 (again with logarithmic factors of s ignored), as can be explicitly
verified by a calculation.

9.3 $f + V \rightarrow f + V$: Sixth Order

At this point one may begin to wonder if the scattering amplitude for
Compton scattering is, at most, of the order of s^0 in all orders of perturbation. A moment's reflection easily convinces us that this cannot be true. A
vector meson may turn into a fermion-antifermion pair part of the time.
Thus fermion–vector meson scattering can be considered as the scattering
of a fermion from a fermion-antifermion pair. Therefore the fermion–vector
meson scattering amplitude at high energies must be equal to the product
of the fermion-fermion scattering amplitude times the fermion-antifermion
scattering amplitude integrated over the wave functions of the fermion-
antifermion state.

These considerations are indeed borne out in a study of sixth-order
diagrams. To this order, only the three diagrams illustrated in figure 9.3 are
important, the other diagrams contributing only s^0 terms with possible
additional factors of $\ln s$. In these three diagrams the incoming vector
meson turns into a pair that exchanges vector mesons with the incoming
fermion and annihilates into the outgoing vector meson. The scattering
amplitude is

$$\mathcal{M}(s, t) = \mathcal{M}_a(s, t) + \mathcal{M}_b(s, t) + \mathcal{M}_c(s, t), \tag{9.26}$$

where the amplitudes on the right-hand side correspond to figures 9.3a,
9.3b, and 9.3c, respectively. Applying the Feynman rules, we get

$$\mathcal{M}_a(s, t) = -2g^6 \int \frac{d^4q}{(2\pi)^4} \frac{[\gamma^\nu(\not{p}_1 - \not{q} + m)\gamma^\mu] A_{ij\nu\mu}(k_1, k_2, q)}{[(p_1 - q)^2 - m^2](q^2 - \lambda^2)[(\Delta - q)^2 - \lambda^2]}, \tag{9.27}$$

$$\mathcal{M}_b(s, t) = -2g^6 \int \frac{d^4q}{(2\pi)^4} \frac{[\gamma^\mu(\not{p}_2 + \not{q} + m)\gamma^\nu] A_{ij\nu\mu}(k_1, k_2, q)}{[(p_2 + q)^2 - m^2](q^2 - \lambda^2)[(\Delta - q)^2 - \lambda^2]}, \tag{9.28}$$

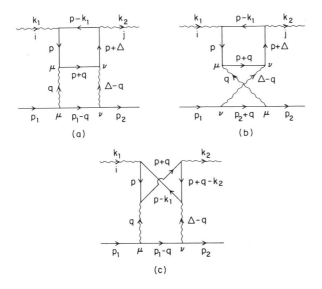

Figure 9.3
The dominant sixth-order diagrams for fermion–vector meson elastic scattering.

$$\mathcal{M}_c(s,t) = -2g^6 \int \frac{d^4q}{(2\pi)^4} \frac{[\gamma^\nu(\not{p}_1 - \not{q} + m)\gamma^\mu]B_{ivj\mu}(k_1,k_2,q)}{[(p_1-q)^2-m^2](q^2-\lambda^2)[(\Delta-q)^2-\lambda^2]}, \quad (9.29)$$

where

$$A_{ijv\mu}$$

$$= \int \frac{d^4p}{(2\pi)^4} \frac{\text{Tr}[\gamma_i(\not{p}+m)\gamma_\mu(\not{p}+\not{q}+m)\gamma_\nu(\not{p}+\not{\Delta}+m)\gamma_j(\not{p}-\not{k}_1+m)]}{(p^2-m^2)[(p+q)^2-m^2][(p+\Delta)^2-m^2][(p-k_1)^2-m^2]}, \quad (9.30)$$

$$B_{ivj\mu} = \int \frac{d^4p}{(2\pi)^4}$$

$$\times \frac{\text{Tr}[\gamma_i(\not{p}+m)\gamma_\mu(\not{p}+\not{q}+m)\gamma_j(\not{p}+\not{q}-\not{k}_2+m)\gamma_\nu(\not{p}-\not{k}_1+m)]}{(p^2-m^2)[(p+q)^2-m^2][(p+q-k_2)^2-m^2][(p-k_1)^2-m^2]}.$$

$$(9.31)$$

In equations (9.27) through (9.29) the numerator is understood to be inserted between the electron spinor wave functions, and the factor of 2 takes into account that there are two possible directions to draw an electron loop. The functions A and B are essentially the fourth-order photon–photon scattering amplitudes. The function B satisfies

$$B_{iv j\mu}(k_1, k_2, q) = B_{i\mu jv}(k_1, k_2, \Delta - q). \tag{9.32}$$

Equation (9.32) can be proved by the change of variable

$$p \to k_1 - p.$$

By substituting equations (9.27) through (9.29) into equation (9.26) and by making use of equation (9.32), we get

$$\mathcal{M}(s,t) = -g^6 \int \frac{d^4 q}{(2\pi)^4} \left[\frac{\gamma^v (\not{p}_1 - \not{q} + m)\gamma^\mu}{(p_1 - q)^2 - m^2} + \frac{\gamma^\mu (\not{p}_2 + \not{q} + m)\gamma^v}{(p_2 + q)^2 - m^2} \right]$$

$$\times \frac{[A_{ijv\mu}(k_1, k_2, q) + A_{ij\mu v}(k_1, k_2, \Delta - q) + B_{iv j\mu}(k_1, k_2, q)]}{(q^2 - \lambda^2)[(\Delta - q)^2 - \lambda^2]}. \tag{9.33}$$

We calculate $\mathcal{M}$ in the center-of-mass frame in which the incident momenta are along the z-axis, with

$$p_{1-} \sim k_{1+} \sim 2\omega \gg m.$$

We then have

$$\frac{\gamma_v (\not{p}_1 - \not{q} + m)\gamma_\mu}{(p_1 - q)^2 - m^2} + \frac{\gamma_\mu (\not{p}_2 + \not{q} + m)\gamma_v}{(p_2 + q)^2 - m^2} \sim -\frac{p_{1\mu} p_{1v}}{m\omega} 2\pi i \delta(q_+), \tag{9.34}$$

where we have made the same approximation as in relations (3.15). With the delta function in relation (9.34), the vector meson propagators are independent of q_+ and q_-; that is,

$$q^2 - \lambda^2 \simeq -\mathbf{q}_\perp^2 - \lambda^2, \qquad (\Delta - q)^2 - \lambda^2 \simeq -(\Delta - \mathbf{q}_\perp)^2 - \lambda^2.$$

Thus

$$\mathcal{M}(s,t) \sim \frac{1}{2} ig^6 \frac{p_{1\mu} p_{1v}}{m\omega} \int \frac{d^2 q_\perp}{(2\pi)^2} \frac{1}{(\mathbf{q}_\perp^2 + \lambda^2)[(\Delta - \mathbf{q}_\perp)^2 + \lambda^2]}$$

$$\times \int_{-\infty}^{\infty} \frac{dq_-}{2\pi} [A_{ijv\mu}(k_1, k_2, q) + A_{ij\mu v}(k_1, k_2, \Delta - q)$$

$$+ B_{iv j\mu}(k_1, k_2, q)] \Big|_{q_+ = 0}. \tag{9.35}$$

Relation (9.35) is just relation (9.1) for fermion–vector meson scattering if we identify

$$\mathscr{I}^V(\mathbf{\Delta}, \mathbf{q}_\perp) = \mathscr{I}_1^V(\mathbf{\Delta}, \mathbf{q}_\perp) + \mathscr{I}_2^V(\mathbf{\Delta}, \mathbf{q}_\perp), \tag{9.36}$$

where

$$\mathscr{I}_1^V(\mathbf{\Delta}, \mathbf{q}_\perp)$$

$$= \frac{1}{4} g^4 \omega^{-3} \int_{-\infty}^{\infty} \frac{dq_-}{2\pi} p_{1\mu} p_{1\nu} [A_{ij\nu\mu}(k_1, k_2, q) + A_{ij\mu\nu}(k_1, k_2, \Delta - q)], \tag{9.37}$$

$$\mathscr{I}_2^V(\mathbf{\Delta}, \mathbf{q}_\perp) = \frac{1}{4} g^3 \omega^{-3} \int \frac{dq_-}{2\pi} p_{1\mu} p_{1\nu} B_{iv j\mu}(k_1, k_2, q), \tag{9.38}$$

with q_+ set to 0.

We calculate $\mathscr{I}^V(\mathbf{\Delta}, \mathbf{q}_\perp)$ in appendix A and show that it is equal to equation (9.3). In particular, it is independent of ω. Readers who are not interested in mathematical details may skip this calculation without losing continuity. We also point out that there is one lesson in the considerations mentioned. For high-energy scattering amplitudes, the lowest-order diagrams do not necessarily dominate the higher-order diagrams, even if the coupling constant is small. For definiteness, let us consider Compton scattering ($e + \gamma \to e + \gamma$) in QED. Then the second-order amplitude in the high-energy limit is of the order of e^2, and the sixth-order amplitude in the high-energy limit is of the order of $(s/m^2)e^6$. The ratio of these two amplitudes is therefore of the order of

$$\frac{m^2}{s} \frac{1}{\alpha^2}.$$

For $s = 1$ (GeV)^2, the number $(m^2/s)(1/\alpha^2)$ is only about 0.005; that is, the sixth-order amplitude is two orders of magnitude larger than the second-order amplitude.

9.4   Sum of Multi-Vector-Meson Exchanges

In this section we discuss fermion–vector meson scattering with the exchange of an arbitrary number of vector mesons. There are two kinds of diagram, as illustrated in figure 9.4. In both kinds of diagram the incoming vector meson turns into a fermion-antifermion pair, which, after exchanging vector mesons with the incident fermion, combines to turn into the out-

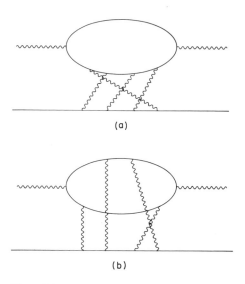

(a)

(b)

Figure 9.4
Two of the diagrams for elastic fermion–vector meson scattering with the exchange of
four vector mesons.

going vector meson. In figure 9.4a, one particle of the pair exchanges an
even number of vector mesons with the incident fermion while the other
particle of the pair propagates freely. (By Furry's theorem [2], the amplitude for the exchange of an odd number of vector mesons vanishes.) In
figure 9.4b, both particles of the pair exchange vector mesons with the
incident fermion.

Let us first calculate the sum of all amplitudes corresponding to the
diagrams of the type in figure 9.4a. The sum of an even number of vector
meson exchanges is proportional to the imaginary part of the right-hand
side of identity (9.6). Thus it is proportional to

$$
\int d^2x_\perp \, e^{i\Delta_\perp \cdot \mathbf{x}_\perp} \left\{ 1 - \cos\left[\frac{g^2}{2\pi} K_0(\lambda|\mathbf{x}_\perp|) \right] \right\}
$$

$$
= \frac{1}{2} \int d^2x_\perp \, e^{i\Delta_\perp \cdot \mathbf{x}_\perp} \left\{ 1 - \exp\left[-i\frac{g^2}{2\pi} K_0(\lambda|\mathbf{x}_\perp|) \right] \right\}
$$

$$
\times \left\{ 1 - \exp\left[i\frac{g^2}{2\pi} K_0(\lambda|\mathbf{x}_\perp|) \right] \right\}. \quad (9.39)
$$

Because the Fourier transform of the product of two functions is equal to the convolution of the Fourier transforms of these two functions, the expression in equation (9.39) is proportional to

$$\int \frac{d^2 q_\perp}{(2\pi)^2} P_-(\mathbf{q}_\perp) P_+(\boldsymbol{\Delta} - \mathbf{q}_\perp). \tag{9.40}$$

The lowest-order term of expression (9.40) is

$$\int \frac{d^2 q_\perp}{(2\pi)^2} \frac{1}{(\mathbf{q}_\perp^2 + \lambda^2)[(\boldsymbol{\Delta} - \mathbf{q}_\perp)^2 + \lambda^2]}, \tag{9.41}$$

which is obtained from expression (9.40) by replacing $P_\pm(\mathbf{p}_\perp)$ with $(\mathbf{p}_\perp^2 + \lambda^2)^{-1}$. Now the lowest-order diagrams of the type in figure 9.4a are the two diagrams in figures 9.3a and 9.3b, which give the amplitude (9.1) (with $a = V$ and $b = f$) with $\mathscr{I}^V$ replaced by $\mathscr{I}_1^V$. We can obtain from this amplitude the sum of amplitudes corresponding to all diagrams of the type in figure 9.4a by replacing $(\mathbf{q}_\perp^2 + \lambda^2)^{-1}[(\boldsymbol{\Delta} - \mathbf{q}_\perp)^2 + \lambda^2]^{-1}$ with $P_-(\mathbf{q}_\perp) P_+(\boldsymbol{\Delta} - \mathbf{q}_\perp)$. The result is expression (9.7) with $\mathscr{I}^V$ replaced by $\mathscr{I}_1^V$.

Similarly, the lowest-order diagram of the type in figure 9.4b is the diagram in figure 9.3c, which gives the amplitude (9.1) (with $a = V$ and $b = f$) with $\mathscr{I}^V$ replaced by $\mathscr{I}_2^V$. We can obtain from this amplitude the sum of amplitudes corresponding to all diagrams of the type in figure 9.4b by replacing the propagators of the vector mesons by $P_-(\mathbf{q}_\perp) P_+(\boldsymbol{\Delta} - \mathbf{q}_\perp)$. The result is expression (9.7) with $\mathscr{I}^V$ replaced by $\mathscr{I}_2^V$. Thus the sum of fermion–vector meson scattering amplitudes of multi-vector-meson exchanges is equal to expression (9.7).

The amplitude for $V + V \to V + V$ and the Delbrück amplitude can be derived in a similar way. The results are given by expressions (9.12) and (9.20). Readers who are interested in the details of the derivation are advised to go to the literature [1].

In this chapter we have concentrated on elastic scattering. Actually, with suitable modifications the results can be generalized to inelastic diffractive processes, the important feature being the absence of quantum numbers in the cross channel [14].

References

[1] H. Cheng and T. T. Wu, *Physical Review Letters* 22 (1969):666, *Physical Review* 182 (1969):1852, 1868, 1873, 1899.

[2] W. H. Furry, *Physical Review* 51 (1937):125.

[3] R. J. Glauber, *Physical Review* 91 (1953):459, 100 (1955):242.

[4] H. Cheng and T. T. Wu, *Physical Review D* 6 (1972):2637; T. Jaroszewicz, *Lettere al Nuovo Cimento* 8 (1973):900; J. Gunion and L. Stodolsky, *Physical Review Letters* 30 (1973):345; E. Papp, *International Journal of Theoretical Physics* 15 (1976):735.

[5] F. Low, *Physical Review D* 12 (1975):163.

[6] A. Chodos, R. L. Jaffe, K. Johnson, C. B. Thorn, and V. F. Weisskopf, *Physical Review D* 9 (1974):3471; T. T. Wu, B. M. McCoy, and H. Cheng, *Physical Review D* 9 (1974):3495.

[7] M. Delbrück, *Zeitschrift für Physik* 84 (1933):144. For a review of Delbrück scattering, see P. Papatzacos and K. Mork, *Physics Reports C* 21 (1975):82.

[8] G. Jarlskog, L. Jönsson, S. Prünster, H. D. Schulz, H. J. Willutzki, and G. G. Winter, *Physical Review D* 8 (1973):3813.

[9] V. Costantini, B. DeTollis, and G. Pistoni, *Nuovo Cimento A* 2 (1971):733; E. A. Vinokurov and E. A. Kuraev, *Soviet Physics JETP* 36 (1973):602; J. Kraus, *Nuclear Physics B* 89 (1975):133; P. Papatzacos and K. Mork, *Physical Review D* 12 (1975):206; B. DeTollis and G. Pistoni, *Nuovo Cimento A* 42 (1977):499; P. Meyer, *Physical Review D* 19 (1979):1680.

[10] R. Moreh and S. Kahane, *Physics Letters B* 47 (1973):351; S. Kahane and R. Moreh, *Physical Review C* 9 (1974):2384; T. Erber, D. White, W. Y. Tsai, and H. G. Latal, *Annals of Physics* 102 (1976):405; S. Kahane, T. Bar-Noy, and R. Moreh, *Nuclear Physics A* 280 (1977):180; T. Bar-Noy and R. Moreh, *Nuclear Physics A* 288 (1977):192; S. Kahane, O. Shahal, and R. Moreh, *Physics Letters B* 66 (1977):229; M. Schumacher and P. Rullhusen, *Physics Letters B* 71 (1977):276; S. Kahane and R. Moreh, *Nuclear Physics A* 308 (1978):88; W. Mückenheim and M. Schumacher, *Journal of Physics G* 6 (1980):1237.

[11] H. Cheng and T. T. Wu, *Physical Review D* 1 (1970):3414; R. W. Brown and I. J. Muzinich, *Physical Review D* 4 (1971):1496; R. W. Brown, W. F. Hunt, K. O. Mikaelian, and I. J. Muzinich, *Physical Review D* 8 (1973):3083.

[12] M. Masujima, *Nuclear Physics B* 24 (1970):182.

[13] K. J. Mork, *Physical Review A* 4 (1971):917; K. J. Mork and H. Olsen, *Physical Review B* 140 (1965):1661.

[14] C. A. Nelson, *Physical Review D* 2 (1970):2669; H. Cheng and T. T. Wu, *Physical Review D* 1 (1970):456, 459, *Physics Letters B* 36 (1971):357; F. J. Gilman, J. Pumplin, A. Schwimmer, and L. Stodolsky, *Physics Letters B* 31 (1970):387; J. Pumplin, *Physical Review D* 4 (1971):3482, 7 (1973):795; A. Białas, W. Czyż, and A. Kotański, *Annals of Physics* 73 (1972):439, *Nuclear Physics B* 46 (1972):109; G. V. Dubrovskii and A. V. Bodganov, *Soviet Physics JETP* 37 (1973):801; G. L. Kotkin and V. G. Serbo, *Soviet Journal of Nuclear Physics* 21 (1975):405; V. G. Zima and N. P. Merenkov, *Soviet Journal of Nuclear Physics* 24 (1976):522; B. Humpert, *Physics Letters B* 68 (1977):66, *Physical Review D* 17 (1978):2955.

10 Method of Calculation by means of Momentum Variables

10.1 Introduction

The calculation of high-energy amplitudes has the undeserved reputation of being difficult. In truth, it is simple and straightforward. Indeed, precisely because it is so simple, one is able to calculate Feynman amplitudes of high orders. The expression for a high-order amplitude is of course long, creating the misunderstanding. Actually, with systematic bookkeeping, even a high-order amplitude can be easily calculated.

A Feynman amplitude can be written as an integral over momentum variables. Alternatively, we can introduce Feynman parameters and then carry out the integration over the momentum variables exactly. The Feynman amplitude is then in the form of an integral over the Feynman parameters. The asymptotic behavior of the Feynman amplitude in the high-energy limit can be obtained by making approximations in either of these two forms.

Making approximations is an art that requires care and experience; the lack of either is an invitation to absurd results. Historically, the first high-energy approximations in gauge theories were made in the Feynman parameter formulation. This route is more conventional and rigorous, and the approximations are more under control; however, the procedure is also tedious and long, and the expressions involved in the intermediate steps are usually complicated. That the final expressions are invariably much simpler suggests that a shorter derivation is possible. The approximation in the momentum variables formulation is precisely that.

Such an approximation in the momentum variables was first employed by Sudakov [1] almost three decades ago in his calculation of the vertex function in the limit of large momentum transfer. There are by now many papers using the Sudakov variables [2]. Ten years after Sudakov, Weinberg [3] suggested the use of the infinite-momentum frame in the completely different context of current algebra. He applied perturbation theory in a Lorentz frame with infinite total momentum for problems in which all Mandelstam variables are not large. The fundamental distinction between these two cases is that, whereas Sudakov studied the physical limit of large momentum transfers, Weinberg used the infinite-momentum frame as a mathematical device, no physical limit being involved. In this chapter we use the momentum variables to study the high-energy behavior of scattering processes in gauge field theories [4]. This treatment also involves a physical limit, namely, that of high energies with fixed momentum transfers, and is thus closer in spirit to that of Sudakov. As already discussed in chapter 5, the fundamental difference between our problem and that of Sudakov is that for our scattering problems the major contributions come from the region where internal lines are not far away from the mass shell. It is this property that underlies the success of the present approach. Various versions of this method of momentum integration for scattering processes have been employed by many authors [5].

In this chapter we discuss the approximation in the momentum variables formulation. Other schemes of approximation have also been developed [6]. A discussion of the approximation in the Feynman parameters formulation is given in appendix C [7].

We choose the Lorentz frame in such a way that the spatial momenta of the two incident particles are in the positive z direction and the negative z direction, with magnitudes ω and ω', respectively. The system in which $\omega = \omega'$ is the center-of-mass system; the system in which $\omega' = 0$ is the laboratory system; and the system in which $\omega = 0$ is the projectile system. These three systems are related by Lorentz boosts in the z direction. A general Lorentz boost in the z direction is most conveniently expressed in the plus, the minus, and the transverse components. Let p_μ and p'_μ be the components of a four-vector in the two Lorentz frames related by a Lorentz boost in the z direction with the relative velocity v; then we have

$$p'_+ = \sqrt{\frac{1+v}{1-v}}\, p_+,$$

$$p'_- = \sqrt{\frac{1-v}{1+v}}\, p_-,$$

$$\mathbf{p}'_\perp = \mathbf{p}_\perp. \tag{10.1}$$

To be specific, we assume that the center-of-mass frame is used, although the calculations can be carried out with equal ease in any Lorentz frame related to the center-of-mass frame by a Lorentz boost in the z direction.

The main idea of calculation is to perform explicitly the integration over all plus momentum variables by contour integration. The integration over the minus momentum variables can then be carried out almost trivially with suitable approximations. The latter integration may be divergent if we set s to be infinite. The divergence is always logarithmic (sometimes an individual diagram may have a divergence that is more than logarithmic, but in gauge field theories such strong divergences invariably cancel as we sum over all diagrams of a given perturbative order), and such a divergent integral is easily identified to be a power of ln s. The asymptotic form of a Feynman amplitude is therefore proportional to a power of ln s, with the coefficient of proportionality equal to an integral over variables of transverse momenta. Such integrals over transverse momenta remain convergent as $s \to \infty$ (again, sometimes an individual diagram may yield a divergent transverse momentum integral, but in gauge field theories such divergences invariably cancel as we sum over all diagrams of a given perturbative order). In other words, these integrals approach constants as $s \to \infty$.

10.2   The Box Diagram

We demonstrate the method of approximation by the simple example of the box diagram in figure 10.1. The solid lines in the figure represent scalar mesons. The box diagram has only one closed loop with the loop momentum designated by q. We carry out the integration of $q_+ = q_0 + q_3$ by contour integration. To do this, we first examine the singularity of the integrand of the Feynman integral. Take, for example, the propagator with momentum $p_1 + q$. The denominator for this propagator is

$$(p_{1+} + q_+)(p_{1-} + q_-) - \mathbf{q}_\perp^2 - \mu^2 + i\varepsilon,$$

which is *linear* in q_+. Thus this propagator has exactly one pole located at

$$q_+ = -p_{1+} + \frac{\mathbf{q}_\perp^2 + \mu^2 - i\varepsilon}{(p_{1-} + q_-)}. \tag{10.2}$$

The important point to note in equation (10.2) is that the pole is located in the lower-half (resp. upper-half) q_+-plane if $p_{1-} + q_-$, the minus momentum of the line, is positive (resp. negative).

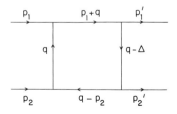

Figure 10.1
The box diagram for the elastic scattering of two scalar mesons. All lines represent scalar mesons of mass μ.

In a Feynman diagram there is no restriction on the values of the minus momenta of the virtual lines so long as momentum conservation is obeyed at each vertex. Consider again the line of momentum q in figure 10.1. The minus component of this line, q_-, may be either positive or negative. To represent explicitly these two possibilities, we draw the two diagrams of figures 10.2a and 10.2b. These two diagrams are called *flow diagrams*, each of which represents a separate kinematic region of q_-. The kinematic region for figure 10.2a is restricted to $q_- > 0$, whereas that for figure 10.2b is restricted to $q_- < 0$, as indicated by the direction of the arrows on line 1 of these figures. Because the direction of the arrow on line 1 in figure 10.2b is opposite to that in figure 10.2a, we designate the momentum of line 1 in figure 10.2b as q' (instead of $-q$). Thus $q'_- > 0$ in this figure.

Similarly, the minus component of lines 2, 3, and 4 may be either positive or negative. To represent explicitly these possibilities, more flow diagrams must be drawn. We make the approximation of setting p_{1-} and p'_{1-} to 0. (This approximation is justified only in the calculation of the leading logarithms.) Then the conservation of the minus momentum at the upper vertices of the diagram dictates that an arrow does not change its direction at either of these vertices. Thus the directions of the arrows on lines 2 and 3 must be the same as the direction of the arrow on line 1. Finally, the arrow on line 4 may be drawn in either direction. To accommodate this,

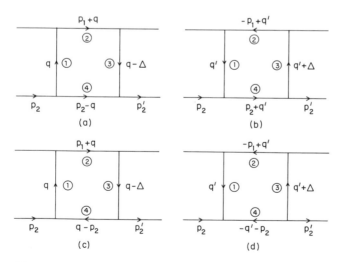

Figure 10.2
The flow diagrams corresponding to the box diagram of figure 10.1.

we draw the flow diagram in figure 10.2c, which differs from figure 10.2a only in the direction of the arrow on line 4. Thus figure 10.2a represents the case $p_{2-} - q_- > 0$, and figure 10.2c represents the case $q_- - p_{2-} > 0$. Similarly, the direction of the arrow on line 4 in figure 10.2b represents the case $q'_- + p_{2-} > 0$. Indeed, the minus momentum of any line in a flow diagram is by definition positive. The flow diagram in figure 10.2d, with the arrow on line 4 pointing in the opposite direction to that in figure 10.2b, is actually not allowed. This is because the conservation of minus momentum cannot be satisfied at either of the lower vertices (the three arrows at a lower vertex are all pointing either inward or outward). Mathematically, it says that $q'_- + p_{2-}$ cannot be negative, as q'_- and p_{2-} are both positive.

In figure 10.2b, the directions of the arrows on all lines of the internal loop are the same. This means that the momentum of any internal line in this figure is of the form of $l + q'$ (that is, none of them is of the form $l - q'$), where l is a linear combination of the external momenta. Thus the poles of the integrand are all located in the lower-half q'_+-plane. By closing the contour in the upper half-plane, we find that the contour integral over q'_+ vanishes. Such a kinematic region does not contribute to the scattering amplitude and can be disregarded. Similar consideration applies to figure 10.2c. Therefore the only contributing flow diagram is figure 10.2a.

Let us now calculate the contribution of the flow diagram in figure 10.2a. We note from the directions of the arrows that the poles from the propagators of lines 1, 2, and 3 lie on the same side of the real q_+-axis, whereas the pole of line 4 lies on the other side. We can close the contour to enclose either the first three poles, or the fourth one. It is obviously simpler to enclose one pole. We redraw the flow diagram in figure 10.2a as figure 10.3 with a cross on line 4 to remind us that the pole of this line is chosen.

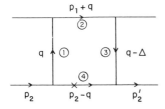

Figure 10.3
The only flow diagram that corresponds to the box diagram and yields a nonzero amplitude in the high-energy approximation. The cross indicates the pole that is enclosed in the contour integration.

Let

$$q_- = 2\omega x. \tag{10.3}$$

Then the arrow on line 1 specifies that $x > 0$, and the arrow on line 4 specifies that $x < 1$. Thus

$$0 < x < 1. \tag{10.4}$$

We define

$$a_i \equiv \mathbf{k}_{i\perp}^2 + \mu_i^2,$$

where k_i and μ_i are the momentum and the mass, respectively, of the ith line. The quantity a_i is known as the square of the transverse mass. Then we have at the pole of the ith line

$$k_{i+} = a_i/k_{i-}. \tag{10.5}$$

Consistent with the approximation of setting $p_{1-} = 0$, we also set $p_{2+} = 0$. Then at the pole of line 4, we have, according to equation (10.5),

$$-q_+ = \frac{a_4}{2\omega(1 - x)}. \tag{10.6}$$

At this pole the values of the denominators for lines 1, 2, and 3 are, respectively,

$$D_1 = (2\omega x)\left[-\frac{a_4}{2\omega(1 - x)} \right] - a_1 = -\frac{xa_4}{1 - x} - a_1,$$

$$D_2 = (2\omega)(2\omega x) - a_2 \simeq sx,$$

$$D_3 = (2\omega x)\left[-\frac{a_4}{2\omega(1 - x)} \right] - a_3 = -\frac{xa_4}{1 - x} - a_3.$$

Thus we have

$$\int_{-\infty}^{\infty} \frac{dq_+}{\prod_{i=1}^{4}(k_i^2 - \mu_i^2)} = \frac{-2\pi i}{2\omega(1 - x)D_1 D_2 D_3}. \tag{10.7}$$

The factor $2\omega(1 - x)$, the minus momentum of line 4, appears in the answer because the residue of $(aq_+ - b)^{-1}F(q_+)$ at $q_+ = b/a$ is $F(b/a)/a$, and the factor $(-2\pi i)$ comes from the Cauchy residue theorem.

Next we carry out the integration over q_-. We have

$$\frac{dq_0 dq_3}{(2\pi)^2} = \frac{1}{2}\frac{dq_+}{2\pi}\frac{dq_-}{2\pi} = \omega\frac{dq_+}{2\pi}\frac{dx}{2\pi}, \tag{10.8}$$

where x is restricted by relation (10.4). From equations (10.7) and (10.8) we get

$$\int \frac{dq_0 dq_3/(2\pi)^2}{\prod_{i=1}^{4}(k_i^2 - \mu_i^2)} = \frac{-i}{4\pi}\int_0^1 \frac{dx}{(1-x)D_1 D_2 D_3}. \tag{10.9}$$

Now the integral in equation (10.9) is divergent at $x = 0$ (note the factor x in D_2). Thus the integral in equation (10.9) is approximately

$$-\frac{i}{4\pi a_1 a_3 s}\int_0^1 \frac{dx}{x} \rightarrow -\frac{i}{4\pi a_1 a_3}\frac{\ln s}{s}, \tag{10.10}$$

where we have set $x = 0$ in the integrand except in D_2 and made the replacement

$$\int_0^1 \frac{dx}{x} \rightarrow \int_{1/s}^1 \frac{dx}{x} = \ln s. \tag{10.11}$$

The cutoff of the lower limit is necessitated by the fact that our approximation breaks down if q_- is of the order of ω^{-1}, as we have approximated p_- (of the order of ω^{-1}) to be 0. From equation (10.3) the cutoff of x is therefore of the order of s^{-1}. If we replace the cutoff by a/s, then the right-hand side of equation (10.11) is $(\ln s - \ln a)$, which differs from $\ln s$ only by an inconsequential real constant.

We therefore obtain for the amplitude of the box diagrams

$$\mathcal{M} = -ig^4 \int \frac{d^4 q}{(2\pi)^4}\frac{1}{\prod_{i=1}^{4}(k_i^2 - \mu_i^2)} \simeq -\frac{g^4}{2s}\frac{\ln s}{2\pi}\int \frac{d^2 q_\perp}{(2\pi)^2}\frac{1}{a_1 a_3}. \tag{10.12}$$

The expression in the right-hand side of equation (10.12) is real. If one desires to obtain the leading imaginary part of the scattering amplitude, one need only observe that there is a term $i\varepsilon$ in the denominator of the propagator. Thus D_2 is really approximately

$$D_2 \simeq sx - a_2 + i\varepsilon$$

and vanishes at $x = a_2/s$. Thus expression (10.11) should be replaced by

$$\int_0^1 \frac{dx}{x} \rightarrow \int_{1/s}^1 \frac{dx}{x} - i\pi = \ln(se^{-i\pi}).$$

Note that the imaginary part is smaller than the real part by a factor of $\ln s$. Thus equation (10.12) becomes

$$\mathcal{M} \simeq -\frac{g^4}{2s} \frac{\ln(se^{-i\pi})}{2\pi} \int \frac{d^2 q_\perp}{(2\pi)^2} \frac{1}{a_1 a_3}. \tag{10.13}$$

Relation (10.13) gives both the leading real part and the leading imaginary part of the scattering amplitude for the box diagram.

The amplitude for the diagram in figure 10.4, with the solid lines representing fermions and the wavy lines representing vector mesons, has the additional numerator factor

$$[\gamma_\nu(\not{p}_1 + \not{q} + m)\gamma_\mu][\gamma^\nu(\not{p}_2 - \not{q} + m)\gamma^\mu] \simeq (\gamma_\nu \not{p}_1 \gamma_\mu)(\gamma^\nu \not{p}_2 \gamma^\mu) \simeq s^2/m^2,$$

as in relations (3.15). Thus the scattering amplitude corresponding to figure 10.4 is equal to

$$-\frac{g^4 s}{2m^2} \frac{\ln(se^{-i\pi})}{2\pi} \int \frac{d^2 q_\perp}{(2\pi)^2} \frac{1}{(q_\perp^2 + \lambda^2)[(\Delta - q_\perp)^2 + \lambda^2]}, \tag{10.14}$$

with both the leading real part and the leading imaginary part of the amplitude included.

Before closing, let us emphasize that there is an exact analogy between a flow diagram and a circuit of electrical current. First, the minus momentum in a flow diagram is always of positive value, as is the current in a dc circuit. The conservation of minus momentum at each vertex and the

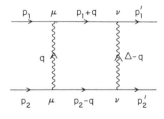

Figure 10.4
The box diagram for fermion-fermion scattering. The wavy lines represent vector mesons, and the solid lines represent fermions.

fact that a flow diagram having a loop with all arrows in the same direction is forbidden are simply Kirchoff's laws. Thus we can draw from the experience of electrical circuitry to determine the flow diagrams. For example, for box diagrams we may immediately recognize that there are only two current outlets: the lower two vertices (no minus momentum comes in or goes out at the upper vertices). The current comes in at the lower left vertex and flows out at the lower right vertex. Thus the voltage at the lower left vertex must be the highest, and the voltage at the lower right vertex must be the lowest. Because a current flows from a point of high voltage to a point of low voltage, the directions of the arrows on lines 1, 3, and 4 must be the way they are drawn in figure 10.3. The direction of the arrow on line 2 is determined from current conservation at either of the two upper vertices. In this way we quickly decide that there is only one flow diagram for the box diagram.

10.3   Other Examples

Let us summarize the procedure of calculations as follows:

1. Draw all possible flow diagrams. Indicate the lines whose poles are enclosed.

2. Carry out the integration over all the plus momenta by calculating the residue of the integrand.

3. Carry out all integrations over the minus momenta. There are possible divergences at the endpoint ($x = 0$ or $x = 1$), which can be identified as logarithmic factors of s.

The rules for drawing the flow diagrams are:

1. There is an arrow on each line of a Feynman diagram. The minus momentum designated in the direction of the arrow is positive.

2. Momentum is conserved at each vertex. Hence the lines on a vertex cannot all point inward or outward.

3. The arrows on the lines in a loop cannot all be in the same direction.

4. If we can ignore the minus momentum of an external line (such as the minus momenta of p_1 and p_1'), the arrow at the vertex involving this external line does not change direction.

5. If the minus components of p_1 and p_1' are ignored, then all arrows on the internal lines on the lower left (resp. lower right) vertex point away from (resp. toward) this vertex.

Ladder Diagrams

Consider the Feynman diagram in figure 10.5a, in which each line represents a scalar meson of mass μ. Let us draw the associated flow diagrams. It is obvious that the arrows on lines 3, 5, and 7 must be the way they are drawn in figure 10.5b, as the left (resp. right) lower vertex represents the point of highest (resp. lowest) voltage in the corresponding electrical circuit. Next, consider the box composed of lines 1, 6, 2, and 4. From our previous discussion of the box diagram, we conclude that there is only one arrow configuration possible—the one drawn in figure 10.5b. Thus the Feynman diagram in figure 10.5a has the diagram in figure 10.5b as its only flow diagram.

Next, let us carry out the integration over the plus components of the loop momenta. There are two loops, and hence we have to carry out two integrations. This can be done simply in the following way. Consider first the loop composed of lines 5, 2, 7, and 3. For this loop we enclose the pole of line 3, which we indicate by putting a cross on it. The only remaining closed loop that does not involve line 3 is the one composed of lines 4, 1, 6, and 2, and we can choose to enclose the pole on line 2, as indicated by the cross in the figure. Thus we have, using equation (10.5),

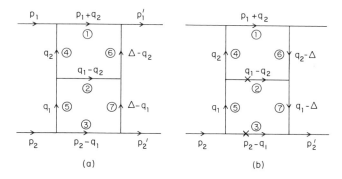

(a)   (b)

Figure 10.5
The sixth-order ladder diagram for the elastic scattering of two scalar mesons: (a) Feynman diagram. (b) Flow diagram. The solid lines represent scalar mesons.

$$-q_{1+} = \frac{a_3}{2\omega(1 - x_1)}, \tag{10.15}$$

and

$$q_{1+} - q_{2+} = \frac{a_2}{2\omega(x_1 - x_2)}, \tag{10.16}$$

where

$$q_{1-} = 2\omega x_1, \tag{10.17}$$

$$q_{2-} = 2\omega x_2, \tag{10.18}$$

with

$$0 < x_2 < x_1 < 1. \tag{10.19}$$

(We conclude that $x_2 < x_1$ from the arrow's direction on line 2.)
We calculate the residue of

$$\prod_{i=1}^{7} (k_i^2 - \mu^2)^{-1}$$

at the poles (10.15) and (10.16) by writing the contribution of each line in successive order. Let

$$\int \frac{\prod_{i=1}^{2} dq_{i+}}{\prod_{i=1}^{7} (k_i^2 - \mu^2)} = \frac{(-2\pi i)^2}{D}. \tag{10.20}$$

Then

$$D \simeq (sx_2)[2\omega(x_1 - x_2)][2\omega(1 - x_1)]\left[-x_2\left(\frac{a_3}{1 - x_1} + \frac{a_2}{x_1 - x_2}\right) - a_4\right]$$

$$\times \left(-x_1\frac{a_3}{1 - x_1} - a_5\right)\left[-x_2\left(\frac{a_3}{1 - x_1} + \frac{a_2}{x_1 - x_2}\right) - a_6\right]$$

$$\times \left(-x_1\frac{a_3}{1 - x_1} - a_7\right), \tag{10.21}$$

where the nth factor in relation (10.21) represents the contribution of line n. The second and the third factors in relation (10.21) are the minus momenta of line 2 and line 3, respectively, the poles of which are enclosed.

The right-hand side of relation (10.21) vanishes at $x_1 = x_2 = 0$ (note the first two factors in relation (10.21)). In particular, in the region

$$1 \gg x_1 \gg x_2,\tag{10.22a}$$

we have

$$D \simeq s^2 x_1 x_2 a_4 a_5 a_6 a_7.\tag{10.22b}$$

Such a region yields two logarithmic factors:

$$\int_{1/s}^{1} dx_1 \int_{1/s}^{x_1} \frac{dx_2}{x_1 x_2} = \frac{1}{2}\ln^2 s.\tag{10.23a}$$

The region where x_2 is comparable to x_1 does not contribute to the $\ln^2 s$ term. This can be checked explicitly by calculating

$$\int_{1/s}^{1} dx_1 \int_{ax_1}^{bx_1} \frac{dx_2}{x_1 x_2} = (\ln b/a)(\ln s).$$

Two logarithmic factors are obtained in regions such as that defined by relation (10.22a) if D is proportional to $x_1 x_2$ in this region.

The leading term in equation (10.23a) is real. If we want to calculate the leading imaginary part as well, we notice that the factor x_2 in relation (10.21) comes from line 1. Just as in the discussion of the box diagram, x_2^{-1} should be replaced by $x_2^{-1} - i\pi\delta(x_2)$. Thus equation (10.23a) should be replaced by

$$\int_{1/s}^{1} dx_1 \int_{1/s}^{x_1} \frac{dx_2}{x_1 x_2} - i\pi \int_{1/s}^{1} \frac{dx_1}{x_1} \simeq \frac{1}{2}\ln^2 (se^{-i\pi}).\tag{10.23b}$$

We therefore have

$$g^6 \int \prod_{i=1}^{2} \frac{d^4 q_i}{(2\pi)^4} \frac{1}{\prod_{i=1}^{7} (k_i^2 - \mu^2)} \simeq -g^6 \int \prod_{i=1}^{2} \frac{d^2 q_{i\perp}}{(2\pi)^2} \int_{1/s}^{1} \frac{\omega\, dx_1}{2\pi} \int_{1/s}^{x_1} \frac{\omega\, dx_2}{2\pi} \frac{1}{D}$$

$$\simeq -\frac{g^6}{8s} \left[\frac{\ln (se^{-i\pi})}{2\pi}\right]^2 \int \prod_{i=1}^{2} \frac{d^2 q_{i\perp}}{(2\pi)^2} \frac{1}{a_4 a_5 a_6 a_7}.\tag{10.24a}$$

The amplitude corresponding to the diagram in figure 10.5 has, in addition, the following factors from the Feynman rules:

$$(-i)(-i)^6(i)^7 = 1,$$

which is the product of the overall factor $(-i)$, a factor $(-i)$ for each vertex, and a factor i for each boson propagator. Thus relation (10.24a) is the asymptotic form for the amplitude of figure 10.5a. Because $a_4 a_6$ (resp. $a_5 a_7$) does not depend on $\mathbf{q}_{1\perp}$ (resp. $\mathbf{q}_{2\perp}$), the integral in relation (10.24a) can be broken up into the product of two integrals. Thus the amplitude corresponding to the ladder diagram (figure 10.5) asymptotically approaches

$$\frac{g^6}{8s} \left[\frac{\ln(se^{-i\pi})}{2\pi} \int \frac{d^2 q_{1\perp}}{(2\pi)^2} \frac{1}{a_5 a_7} \right]^2, \qquad (10.24b)$$

which includes both the leading real part and the leading imaginary part.

Three-Meson Exchange Diagrams

There are six three-meson exchange diagrams. They are illustrated in figure 10.6, where all of the lines represent scalar mesons. Not all the values of the scattering amplitudes are independent. For example, figures 10.6b and 10.6c can be obtained from each other by a right-left reflection ($p_i \leftrightarrow -p_i'$, $i = 1, 2$). Because none of the Mandelstam variables change under such a transformation, we have

$$\mathscr{M}_b = \mathscr{M}_c, \qquad (10.25)$$

where $\mathscr{M}_b$ and $\mathscr{M}_c$ refer to the amplitudes corresponding to figures 10.6b and 10.6c, respectively. Similarly, under the transformation $s \leftrightarrow u$, we have

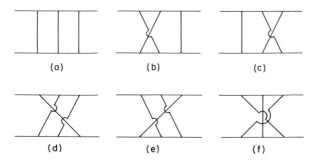

(a)   (b)   (c)

(d)   (e)   (f)

Figure 10.6
Sixth-order diagrams of three-meson exchange. The solid lines represent scalar mesons.

$$\mathcal{M}_a \leftrightarrow \mathcal{M}_f, \tag{10.26}$$

$$\mathcal{M}_b \leftrightarrow \mathcal{M}_e, \tag{10.27}$$

$$\mathcal{M}_c \leftrightarrow \mathcal{M}_d. \tag{10.28}$$

Thus we need to calculate only $\mathcal{M}_a$ and $\mathcal{M}_b$. Consider first the Feynman diagram in figure 10.6a. We draw its flow diagrams in figure 10.7. The directions of the arrows for lines 1, 6, 7, and 3 are easily determined from the fact that the left (resp. right) lower vertex represents the point of the highest (resp. lowest) voltage. The directions of the arrows on lines 4 and 5 are determined from current conservation at the upper vertices. Only the arrow on line 2 can take either direction, and hence there are two flow diagrams (figure 10.7a and figure 10.7b) for the Feynman diagram in figure 10.6a.

Let us begin with the diagram in figure 10.7a. For the loop with lines 1, 4, 2, and 6 we enclose the pole of line 6. For the loop with lines 2, 5, 3, and 7, the arrows on lines 5 and 3 are in one direction and the arrows on lines 2 and 7 are in the other direction. Thus we have to enclose two poles either way. Let us enclose the poles of line 3 and line 5, which we indicate by putting a bar on each of these lines.

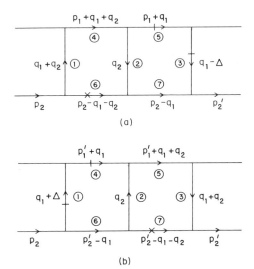

Figure 10.7
The flow diagrams corresponding to figure 10.6a.

To integrate over the first loop, we evaluate the residue at the pole of line 6. At this pole

$$-q_{1+} - q_{2+} \simeq \frac{a_6}{2\omega(1 - x_1 - x_2)},$$
(10.29)

where

$$q_{i-} = 2\omega x_i, \qquad i = 1, 2.$$
(10.30)

We note from the arrows on line 3 and line 2 that

$$x_i > 0, \qquad i = 1, 2,$$

and from the arrow on line 6 that

$$x_1 + x_2 < 1.$$

To integrate over the second loop, we add the residue at pole 3 and the residue at pole 5. Thus the double integral over $dq_{1+} dq_{2+}$ is equal to the sum of two terms: the residue at pole 6 for the first loop and at pole 3 for the second loop, and the residue at pole 6 for the first loop and at pole 5 for the second loop. Let us calculate the first residue. At pole 3

$$q_{1+} = a_3/q_{1-}.$$
(10.31)

The contribution from pole 3 and pole 6 is then given by D^{-1}, where

$$D = \left[-\frac{(x_1 + x_2)a_6}{1 - x_1 - x_2} - a_1 \right]\left[-x_2\left(\frac{a_6}{1 - x_1 - x_2} + \frac{a_3}{x_1} \right) - a_2 \right](2\omega x_1)$$

$$\times [s(x_1 + x_2)](sx_1)[2\omega(1 - x_1 - x_2)]\left[-(1 - x_1)\frac{a_3}{x_1} - a_7 \right].$$
(10.32a)

The dominant contribution comes from the region where x_1 and x_2 are small. In this region

$$D \simeq -s^3 a_1 a_3 (x_2 a_3 + x_1 a_2)(x_1 + x_2).$$
(10.32b)

Note that D is not proportional to $x_1 x_2$ in either the region $x_1 \gg x_2$ or $x_1 \ll x_2$. Thus the amplitude has at most one logarithmic factor of s. The contribution of the set of poles of lines 6 and 3 in the flow diagram of figure 10.7a is then equal to

$$g^6(-i)^2 \int \prod_{i=1}^{2} \frac{d^2 q_{i\perp}}{(2\pi)^2} \int_{1/s}^{1} \frac{\omega \, dx_1}{2\pi} \int_{1/s}^{1} \frac{\omega \, dx_2}{2\pi} \frac{-1}{s^3 a_1 a_3 (x_2 a_3 + x_1 a_2)(x_1 + x_2)}.$$
$$\text{(10.33)}$$

In expression (10.33) each factor of $(-i)$ comes from an integration of the plus component. Let us make the change of variables

$$x_1 = \rho \alpha,$$

$$x_2 = \rho(1 - \alpha).$$

Then

$$dx_1 \, dx_2 = \rho \, d\rho \, d\alpha.$$

It is easy to carry out the integration over ρ:

$$\int_{1/s}^{1} \frac{d\rho}{\rho} = \ln s.$$

Thus the contribution of pole 3 and pole 6 in the flow diagram in figure 10.7a is approximately

$$\frac{g^6 \ln s}{16\pi^2 s^2} \int \prod_{i=1}^{2} \frac{d^2 q_{i\perp}}{(2\pi)^2} \int_{0}^{1} \frac{d\alpha}{a_1 a_3 [\alpha a_2 + (1 - \alpha) a_3]},$$

which is equal to

$$\frac{g^6 \ln s}{16\pi^2 s^2} \int \prod_{i=1}^{2} \frac{d^2 q_{i\perp}}{(2\pi)^2} \frac{\ln(a_2/a_3)}{a_1 a_3 (a_2 - a_3)}, \tag{10.34}$$

which has only one factor of $\ln s$.

Next we calculate the contribution of the set of poles of line 6 and line 5 in the flow diagram in figure 10.7a. We show that this contribution is negligible. At the pole of line 5 we have

$$2\omega + q_{1+} = \frac{a_5}{2\omega x_1}. \tag{10.35a}$$

From equation (10.35a) and relation (10.29) we get

$$q_{2+} = 2\omega - \frac{a_5}{2\omega x_1} - \frac{a_6}{2\omega(1 - x_1 - x_2)}. \tag{10.35b}$$

The contribution with the poles of the two loops taken at line 6 and line 5 is D^{-1}, where

$$D \simeq \left[-\frac{(x_1 + x_2)a_6}{1 - x_1 - x_2} - a_1 \right] (sx_2)(-sx_1)[s(x_1 + x_2)](2\omega x_1)$$

$$\times [2\omega(x_1 + x_2)][(1 - x_1)s]. \tag{10.36}$$

We note that D in relation (10.36) is proportional to s^5, not to s^3 as in the previous case. At first sight this suggests that the contribution from this set of poles is too small by a factor of s^2. Actually, it is smaller only by a factor of $\ln s$. This is because there are more powers of x_1 and x_2 in relation (10.36). We have, for both x_1, x_2 much smaller than unity,

$$D \simeq s^5 x_1^2 x_2 (x_1 + x_2)^2 a_1, \tag{10.37}$$

which has two more powers of x_1 and x_2 than the D function of relation (10.32b). In the region where x_1 and x_2 are of the order of s^{-1}, these two powers of x_1 and x_2 can compensate for the power s^2. Indeed, as we integrate the inverse of relation (10.37) over

$$\int_{1/s}^{1} \frac{\omega \, dx_1}{2\pi} \int_{1/s}^{1} \frac{\omega \, dx_2}{2\pi},$$

we easily find that the value of the integral is of the order of s^{-2}, without any logarithms (for example, put $x_i = sy_i$, $i = 1, 2$). This value is negligible compared to the term in expression (10.34), which is of the order of $s^{-2} \ln s$. Thus expression (10.34) is the contribution of the flow diagram in figure 10.7a to the scattering amplitude.

We can calculate the flow diagram in figure 10.7b in exactly the same way. By using the notation designated in the figure, we see that each of the lines in figure 10.7b can be identified with a line in figure 10.7a, with $p_i' \leftrightarrow p_i$, $i = 1, 2$, and $\Delta \to -\Delta$. Thus the calculation is exactly the same step by step, and we get the same answer as expression (10.34). The amplitude for figure 10.6a is therefore equal to twice the expression (10.34):

$$\mathcal{M}_a \simeq \frac{g^6 \ln s}{8\pi^2 s^2} \int \prod_{i=1}^{2} \frac{d^2 q_{i\perp}}{(2\pi)^2} \frac{\ln(a_2/a_3)}{a_1 a_3 (a_2 - a_3)}. \tag{10.38}$$

We can rewrite relation (10.38) as

$$\mathcal{M}_a \simeq \frac{g^6 \ln s}{8\pi^2 s^2} \int \prod_{i=1}^{3} \frac{d^2 q_{i\perp}}{(2\pi)^2} (2\pi)^2 \delta^{(2)}(\mathbf{q}_{1\perp} + \mathbf{q}_{2\perp} + \mathbf{q}_{3\perp} - \mathbf{\Delta})$$

$$\times \frac{\ln \dfrac{\mathbf{q}_{2\perp}^2 + \mu^2}{\mathbf{q}_{3\perp}^2 + \mu^2}}{(\mathbf{q}_{1\perp}^2 + \mu^2)(\mathbf{q}_{3\perp}^2 + \mu^2)(\mathbf{q}_{2\perp}^2 - \mathbf{q}_{3\perp}^2)}, \quad (10.39)$$

where μ denotes the mass of the vertical lines. This amplitude is smaller than the ladder amplitude (10.24b) by a factor of $\ln s$. The leading imaginary part of $\mathcal{M}_a$ is even smaller, of the order of s^{-2}.

Next we consider the Feynman diagram in figure 10.6b, which has two flow diagrams, illustrated in figure 10.8. Consider first the flow diagram in figure 10.8a. For the loop of lines 1, 6, 3, and 5, we enclose the pole of line 3, and for the loop of lines 1, 5, 4, 7, and 2, we enclose either the pole of line 2 or the pole of line 7. Just as in the preceding example, the contribution of the pole of line 2, an upper horizontal line, is negligible. Thus we need to include only the contribution of the pole of line 7. We have

$$q_{2+} \simeq \frac{a_7}{2\omega x_2}, \qquad q_{1+} \simeq -\frac{a_7}{2\omega x_2}, \qquad (10.40)$$

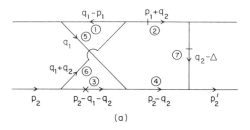

(a)

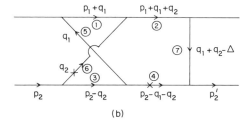

(b)

Figure 10.8
The flow diagrams corresponding to figure 10.6b.

where

$$q_{i-} = 2\omega x_i, \qquad i = 1, 2, \tag{10.41}$$

with

$$0 < x_i < 1, \qquad i = 1, 2.$$

The denominator of the residue is

$$D \simeq (-sx_1)(sx_2)\,[2\omega(1 - x_1 - x_2)]\left[-(1 - x_2)\frac{a_7}{x_2} - a_4\right]$$

$$\times \left(-x_1\frac{a_7}{x_2} - a_5\right)(-a_6)(2\omega x_2). \tag{10.42}$$

For x_1 and x_2 small, D is approximately

$$D \simeq s^3 x_1 x_2 a_7 a_6\left(x_1\frac{a_7}{x_2} + a_5\right). \tag{10.43}$$

In the region $1 \gg x_2 \gg x_1$, D is proportional to $x_1 x_2$. Thus this residue contributes a $\ln^2 s$ term. It turns out, however, that this term is canceled by the leading term of the flow diagram in figure 10.8b. Thus we must calculate the next leading term, which is of the order of $(\ln s)/s^2$. Note that in the region $1 \gg x_2 \gg x_1$ there is no imaginary term of the order of $\ln s$. This is because the factor x_1 comes from line 1 of figure 10.8a, the momentum of which is $q_1 - p_1$, not $q_1 + p_1$. Thus x_1 is really the approximation of

$$(-sx_1 - a_1 + i\varepsilon)/s.$$

Because the zero of this factor is $x_1 = -a_1/s$, which is negative and hence lies outside the region of integration, there is no delta function associated with $1/x_1$, and the contribution of the flow diagram in figure 10.8a to the scattering amplitude is

$$(-i)^2 g^6 \int \prod_{i=1}^{2} \frac{d^2 q_{i\perp}}{(2\pi)^2}\frac{\omega\,dx_i}{2\pi}\frac{1}{s^3 x_1 a_6 a_7(x_1 a_7 + x_2 a_5)}. \tag{10.44}$$

Next we study the flow diagram in figure 10.8b. The integration over q_{1+} and q_{2+} can be carried out by evaluating the residue at the pole of line 6 and the pole of line 4, where

$$q_{1+} \simeq -\frac{a_6}{2\omega x_2}, \qquad q_{2+} = \frac{a_6}{2\omega x_2}. \tag{10.45}$$

The D function for the flow diagram in figure 10.8b is

$$D \simeq (sx_1)\,[s(x_1 + x_2)]\left[-(1 - x_2)\frac{a_6}{x_2} - a_3\right][2\omega(1 - x_1 - x_2)]$$

$$\times \left(-x_1\frac{a_6}{x_2} - a_5\right)(2\omega x_2)(-a_7)$$

$$\simeq -s^3 x_1(x_1 + x_2)\left(x_1\frac{a_6}{x_2} + a_5\right)a_6 a_7, \tag{10.46}$$

where the approximation is valid if both x_1 and x_2 are small. The contribution of the flow diagram in figure 10.8b to the scattering amplitude is hence equal to

$$-(-i)^2 g^6 \int \prod_{i=1}^{2} \frac{d^2 q_{i\perp}}{(2\pi)^2} \frac{\omega\,dx_i}{2\pi} \frac{1}{s^3 x_1(x_1 + x_2)a_6 a_7} \frac{x_2}{x_1 a_6 + x_2 a_5}. \tag{10.47a}$$

We can rename a_6 as a_7 and vice versa in expression (10.47a), as line 6 and line 7 are completely interchangeable if lines 1, 2, 3, and 4 are fused (a_i, $i = 1, 2, 3, 4$, does not appear in the integrand).

The amplitude in expression (10.47a) has again two factors of $\ln s$, which come from the region $x_1 \ll x_2 \ll 1$. This $\ln^2 s$ term just cancels the $\ln^2 s$ term in expression (10.44). Thus the uncanceled leading term has only one factor of $\ln s$. To calculate this uncanceled leading term, we observe that the factor x_1 in relation (10.46) comes from line 1 of figure 10.8b, which is of momentum $(p_1 + q_1)$. Thus x_1 is the approximation of

$$(sx_1 - a_1 - i\varepsilon)/s,$$

and the zero $x_1 = a_1/s$ falls inside the region of integration. Thus expression (10.47a) should be replaced by

$$-\frac{ig^6}{16\pi s^2}\int \prod_{i=1}^{2}\frac{d^2 q_{i\perp}}{(2\pi)^2}\frac{1}{a_5 a_6 a_7}\int \frac{dx_2}{x_2}$$

$$+\frac{g^6}{16\pi^2 s^2}\int \prod_{i=1}^{2}\frac{d^2 q_{i\perp}}{(2\pi)^2}\,dx_i\,\frac{x_2}{x_1(x_1 + x_2)a_6 a_7(x_1 a_7 + x_2 a_5)}, \tag{10.47b}$$

where the first term in expression (10.47b) comes from the delta function of $1/(x_1 + i\varepsilon)$. By adding expression (10.47b) to expression (10.44), we obtain the asymptotic form of the scattering amplitude corresponding to figure 10.6b:

$$\mathcal{M}_b \simeq -\frac{g^6}{16\pi^2 s^2} \int \frac{\prod\limits_{i=1}^{2} [d^2 q_{i\perp} \, dx_i/(2\pi)^2]}{(x_1 + x_2) a_6 a_7 (x_1 a_7 + x_2 a_5)} - \frac{ig^6}{16\pi s^2} \ln s \, I_1, \qquad (10.48)$$

where

$$I_1 = \int \prod_{i=1}^{3} \frac{d^2 q_{i\perp}}{(2\pi)^2} \frac{(2\pi)^2 \delta^{(2)}\left(\sum\limits_{1}^{3} \mathbf{q}_{i\perp} - \mathbf{\Delta}\right)}{\prod\limits_{1}^{3} (\mathbf{q}_{i\perp}^2 + \mu^2)}. \qquad (10.49)$$

The second term in relation (10.48) is just the first term in expression (10.47b). The integral in relation (10.48) diverges when x_1 and x_2 are both small, with x_1 comparable to x_2. Thus, by making the change of variables as in the formulas given after expression (10.33), we obtain the amplitude for the diagrams in figure 10.6b:

$$\mathcal{M}_b \simeq -\frac{g^6 \ln s}{16\pi^2 s^2} (I_2 + i\pi I_1), \qquad (10.50)$$

where

$$I_2 = \int \prod_{i=1}^{2} \frac{d^2 q_{i\perp}}{(2\pi)^2} \frac{\ln(a_7/a_5)}{a_6 a_7 (a_7 - a_5)}. \qquad (10.51)$$

We reiterate that the imaginary part of $\mathcal{M}_b$ is contributed by the region $x_1 \ll x_2 \ll 1$, whereas the real part of $\mathcal{M}_b$ is contributed by the region x_1, $x_2 \ll 1$, with x_1 and x_2 comparable.

We can obtain the rest of the three-meson exchange amplitudes from relations (10.25) through (10.28). For example, for equation (10.25), we get

$$\mathcal{M}_c \simeq -\frac{g^6 \ln s}{16\pi^2 s^2} (I_2 + i\pi I_1). \qquad (10.52)$$

Also, from relation (10.27), $\mathcal{M}_e$ is obtained from $\mathcal{M}_b$ by the replacement of $s \leftrightarrow u$, or $p_1 \leftrightarrow -p_1'$. Now if p_1 is replaced by $-p_1'$, then the momentum in line 1 of the flow diagram in figure 10.8a is $(q_1 + p_1')$. Then the $i\varepsilon$ in the propagator of line 1 should be taken into account. Similarly, we find that the $i\varepsilon$ for line 1 in the flow diagram in figure 10.8b can be ignored. This is opposite to the case of $\mathcal{M}_b$. Thus the imaginary part in relation (10.50) changes sign, and we have

$$\mathcal{M}_d = \mathcal{M}_e \simeq -\frac{g^6 \ln s}{16\pi^2 s^2}(I_2 - i\pi I_1). \tag{10.53}$$

Similarly, we have from relation (10.26) that

$$\mathcal{M}_a \simeq \mathcal{M}_f \simeq \frac{g^6 \ln s}{8\pi^2 s^2}I_2. \tag{10.54}$$

Relations (10.50) through (10.54) give the asymptotic amplitudes for three-meson exchange. All of them are of the order of $(\ln s)/s^2$, a factor of $\ln s$ smaller than the ladder diagram of the sixth order. However, the imaginary parts of $\mathcal{M}_b$, $\mathcal{M}_c$, $\mathcal{M}_d$, and $\mathcal{M}_e$ are comparable to the imaginary part of the sixth-order ladder amplitude.

We note from relations (10.50) through (10.54) that the sum of the six asymptotic amplitudes from the six diagrams in figure 10.6 vanishes. This simply means that this sum is proportional to s^{-2}, not $s^{-2} \ln s$. This is why the sum of three-vector-meson exchange amplitudes discussed in chapter 3 has no logarithmic factors.

An Eighth-Order Diagram

The calculations of the higher-order diagrams are not much more difficult—a virtue of the present method. As an illustration, let us consider the eighth-order Feynman diagram in figure 10.9. The solid lines represent fermions, whereas the wavy lines represent vector mesons. According to the Feynman rules of QED, the integrand of the Feynman integral has a numerator that is of the order of s^4, provided that the plus momentum

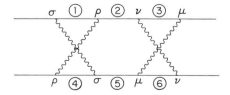

Figure 10.9
An eighth-order diagram for fermion-fermion elastic scattering. The wavy lines represent vector mesons, and the solid lines represent fermions.

passing through lines 1, 2, and 3 and the minus momentum passing through lines 4, 5, and 6 are both of the order of ω. Indeed, so far as the leading term is concerned, the momenta in lines 4, 5, and 6 are all approximately p_2. This means that the arrows on lines 4, 5, and 6 in all flow diagrams point to the right. Also, the momenta passing through lines 1, 2, and 3 are all approximately p_1. Thus there is no contribution from the poles of lines 1, 2, and 3. (Otherwise, the plus momentum of such a line is of the order of ω^{-1}.) Therefore, when we consider the poles to be enclosed, those of lines 1, 2, and 3 need not be taken into account. It then follows that there are only three contributing flow diagrams. These are illustrated in figure 10.10. Furthermore, the numerator of the Feynman integral is approximately

$$N \simeq g^8(\gamma_\mu \not{p}_1 \gamma_\nu \not{p}_1 \gamma_\rho \not{p}_1 \gamma_\sigma)(\gamma^\nu \not{p}_2 \gamma^\mu \not{p}_2 \gamma^\sigma \not{p}_2 \gamma^\rho)$$

$$\simeq 4g^8 s^4/m^2. \tag{10.55}$$

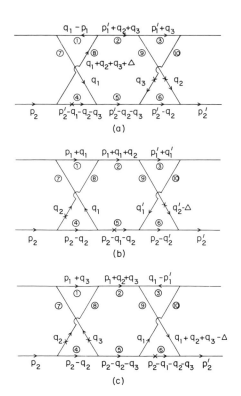

Figure 10.10
The flow diagrams corresponding to figure 10.9.

Consider the flow diagram in figure 10.10a. At the poles indicated by the crosses in the figure, we have

$$q_{2+} = \frac{a_9}{Q_2}, \qquad q_{3+} = \frac{a_{10}}{Q_3}, \qquad q_{1+} \simeq -\frac{a_9}{Q_2} - \frac{a_{10}}{Q_3}, \qquad (10.56)$$

where

$$Q_i \equiv q_{i-}, \qquad i = 1, 2, 3.$$

All Q_i, $i = 1, 2, 3$, are positive, as is evident from the arrows on lines 7, 9, and 10, and range from ω^{-1} to ω. The D function for this flow diagram is

$$D \simeq (-2\omega Q_1)[2\omega(Q_2 + Q_3)](2\omega Q_3)$$

$$\times (2\omega)\left[-2\omega\left(\frac{a_9}{Q_2} + \frac{a_{10}}{Q_3}\right)\right]\left(-2\omega\frac{a_9}{Q_2}\right)$$

$$\times \left[-Q_1\left(\frac{a_9}{Q_2} + \frac{a_{10}}{Q_3}\right) - a_7\right](-a_8)(Q_2)(Q_3), \qquad (10.57)$$

where we have made the approximation $Q_i \ll \omega$, $i = 1, 2, 3$.

The possible regions giving three factors of $\ln s$ are the region $Q_1 \gg Q_2 \gg Q_3$ plus five other regions obtained from it by permuting Q_1, Q_2, and Q_3. We have to check in each region whether D is proportional to $Q_1 Q_2 Q_3$. Take, for example, the region $Q_1 \gg Q_2 \gg Q_3$, where

$$D \propto Q_1^2 Q_2,$$

which is larger than $Q_1 Q_2 Q_3$, as $Q_1 \gg Q_3$. Thus this region does not contribute. We can check all other regions separately, but a clue to the answer is provided by the seventh factor in relation (10.57), which is large if Q_1 is much larger than either Q_2 or Q_3. Thus we must restrict ourselves to the regions in which Q_1 is much smaller than Q_2 and Q_3. The third factor in relation (10.57) suggests further that D is the smallest if Q_3 is smaller than Q_2. Thus the only contributing region is $Q_1 \ll Q_3 \ll Q_2$, where, indeed,

$$D \simeq -s^3 Q_1 Q_2 Q_1 a_7 a_8 a_9 a_{10}. \qquad (10.58)$$

It turns out that the leading real term from the flow diagram in figure 10.10a is canceled by the leading real terms from the flow diagrams in figures 10.10b and 10.10c, but the leading imaginary terms from these flow diagrams do not cancel. The leading imaginary term of order $s\ln^2 s$ for the flow diagram in figure 10.10a, if it exists, comes from the $i\varepsilon$ associated with

Q_1, the smallest of the minus momenta. This is because, with $\alpha > 1$,

$$\text{Im} \int_{s^{-1}}^{1} \frac{dx_2}{x_2} \int_{s^{-1}}^{x_2} \frac{dx_3}{x_3} \int_{s^{-1}}^{x_3} \frac{dx_1}{x_1 - \alpha/s + i\varepsilon} \simeq -\pi \frac{\ln^2 s}{2},$$

whereas

$$\text{Im} \int_{s^{-1}}^{1} \frac{dx_2}{x_2} \int_{s^{-1}}^{1} \frac{dx_3}{x_3 - \alpha/s + i\varepsilon} \int_{s^{-1}}^{1} \frac{dx_1}{x_1} = O(\ln s).$$

Now the factor Q_1 in relation (10.58) comes from line 1, the momentum of which is $(q_1 - p_1)$. Thus the $i\varepsilon$ from this propagator can be dropped, and there is no leading imaginary term of the order of $s \ln^2 s$.

The scattering amplitude contributed by a set of poles after all q_+ integrations have been performed is equal to

$$(-i)(-i)^v(i)^q(-i)^n(-\tfrac{1}{2}i)^l \int \prod_i \frac{d^2 q_{i\perp}}{(2\pi)^2} \frac{dQ_i}{2\pi} \frac{N}{D}. \tag{10.59}$$

In expression (10.59), v is the number of internal lines of vector mesons ($-i$ for each vector meson propagator), q is the number of internal lines of quarks (i for each quark propagator), n is the order of perturbation ($-i$ for each vertex), and l is the number of internal loops ($-\tfrac{1}{2}i$ for each loop). (The factor $\tfrac{1}{2}$ is from the change of variables

$$dq_0 \, dq_3 = \tfrac{1}{2} dq_+ \, dQ,$$

and the factor $-i$ is from the Cauchy residue theorem.) For the flow diagram in figure 10.10a, $v = 4$, $q = 6$, $n = 8$, and $l = 3$. Also, N and D are given by relations (10.55) and (10.58), respectively. Thus the amplitude for the flow diagram in figure 10.10a is

$$\mathcal{M}_a \simeq \frac{1}{8} \int \prod_{i=1}^{3} \frac{d^2 q_{i\perp}}{(2\pi)^2} \frac{dQ_i}{2\pi Q_i} \frac{4g^8 s^4/m^2}{s^3 a_7 a_8 a_9 a_{10}} \simeq \frac{sg^8}{2m^2} \frac{1}{3!} \left(\frac{\ln s}{2\pi}\right)^3 I_3, \tag{10.60}$$

where

$$I_3 = \int \prod_{i=1}^{4} \frac{d^2 q_{i\perp}}{(2\pi)^2} \frac{(2\pi)^2 \delta^{(2)} \left(\sum_1^4 \mathbf{q}_{i\perp} - \boldsymbol{\Delta}\right)}{\prod_1^4 (\mathbf{q}_{i\perp}^2 + \lambda^2)}. \tag{10.61}$$

In deriving equation (10.60) we made use of the formula

$$\int_{1/\omega}^{\omega} \frac{dQ_2}{Q_2} \int_{1/\omega}^{Q_2} \frac{dQ_3}{Q_3} \int_{1/\omega}^{Q_3} \frac{dQ_1}{Q_1} \simeq \frac{\ln^3 s}{3!}.$$

Next we turn to the flow diagram in figure 10.10b. At the poles indicated in the figure we have

$$q'_{2+} \simeq \frac{a_9}{Q'_2}, \qquad q_{2+} = \frac{a_8}{Q_2}, \qquad q_{1+} \simeq -\frac{a_8}{Q_2}. \tag{10.62a}$$

Note that $q_1 + q_2 = q'_1 + q'_2$, and, in particular, that

$$Q_1 + Q_2 = Q'_1 + Q'_2. \tag{10.62b}$$

We have

$$D \simeq (2\omega Q_1)[2\omega(Q_1 + Q_2)](2\omega Q'_1)\left(-2\omega \frac{a_8}{Q_2}\right)(2\omega)\left(-2\omega \frac{a_9}{Q'_2}\right)$$

$$\times \left(-Q_1 \frac{a_8}{Q_2} - a_7\right)(Q_2)(Q'_2)\left(-Q'_1 \frac{a_9}{Q'_2} - a_{10}\right). \tag{10.63}$$

The seventh and tenth factors suggest that the contributing regions must satisfy

$$Q_1 \ll Q_2 \quad \text{and} \quad Q'_1 \ll Q'_2. \tag{10.64}$$

Thus equation (10.62b) gives

$$Q_2 \simeq Q'_2.$$

When the inequalities (10.64) are satisfied, D becomes

$$D \simeq s^3 Q_1 Q'_1 Q_2 a_7 a_8 a_9 a_{10}. \tag{10.65}$$

Thus there are two contributing regions: $Q_1 \ll Q'_1 \ll Q_2$ and $Q'_1 \ll Q_1 \ll Q_2$. The arrows on lines 1 and 3 both point to the right. Thus these lines give rise to leading imaginary terms. We note that relations (10.65) and (10.58) differ only by a sign. Thus the contribution from the flow diagram in figure 10.10b is obtained from equation (10.60) by changing the sign, replacing $\ln s$ by $\ln(se^{-i\pi})$, and multiplying by a factor of 2 (two contributing regions). Therefore

$$\mathcal{M}_b \simeq -\frac{sg^8}{m^2} \frac{1}{3!} \left[\frac{\ln(se^{-i\pi})}{2\pi}\right]^3 I_3. \tag{10.66}$$

Finally, let us consider the flow diagram in figure 10.10c. We observe that, for every internal line in figure 10.10c, there is a corresponding one in figure 10.10a, with the momenta related by the replacement $p_1 \leftrightarrow p_1'$, $p_2 \leftrightarrow p_2'$ (and hence $\Delta \to -\Delta$). (For example, line 6 in figure 10.10c corresponds to line 4 in figure 10.10a.) This is because the flow diagram in figure 10.10c can be obtained from the flow diagram in figure 10.10a by making a right-left reflection and by changing the direction of all the arrows. It is then clear that

$$\mathcal{M}_c \simeq \mathcal{M}_a. \tag{10.67}$$

The leading part of the scattering amplitude corresponding to the Feynman diagram in figure 10.9 is obtained by adding the three amplitudes from relations (10.60), (10.66), and (10.67). We get

$$\mathcal{M} \simeq \mathcal{M}_a + \mathcal{M}_b + \mathcal{M}_c \simeq \frac{is}{4m^2} g^8 \left(\frac{\ln s}{2\pi}\right)^2 I_3. \tag{10.68}$$

References

[1] V. V. Sudakov, *Soviet Physics JETP* 3 (1956): 65.

[2] M. Cassandro and M. Cini, *Nuovo Cimento* 34 (1964): 1719; R. Jackiw, *Annals of Physics* 48 (1968): 292; T. Appelquist and J. R. Primack, *Physical Review D* 1 (1970): 1144; P. M. Fishbane and J. D. Sullivan, *Physical Review D* 4 (1971): 458.

[3] S. Weinberg, *Physical Review* 150 (1966): 1313; S. J. Chang and S. K. Ma, *Physical Review* 180 (1969): 1506.

[4] H. Cheng and T. T. Wu, *Physical Review Letters* 22 (1969): 666, *Physical Review* 182 (1969): 1899.

[5] S. J. Chang and S. K. Ma, *Physical Review Letters* 22 (1969): 1334; J. B. Kogut and D. E. Soper, *Physical Review D* 1 (1970): 2901; J. D. Bjorken, J. B. Kogut, and D. E. Soper, *Physical Review D* 3 (1971): 1382; L. N. Lipatov, and G. V. Frolov, *Soviet Journal of Nuclear Physics* 13 (1971): 333; G. Feldman, and P. T. Matthews, *Journal of Physics A* 6 (1973): 236; T. Garavaglia, *Physical Review D* 12 (1975): 3327.

[6] L. N. Chang and N. P. Chang, *Physical Review D* 4 (1971): 1856; D. R. Harrington, *Physical Review D* 5 (1972): 892; M. A. González, *Physical Review D* 6 (1972): 1756; Q. Bui-Duy, *Physical Review D* 9 (1974): 2794, *Lettere al Nuovo Cimento* 14 (1975): 24.

[7] M. Gell-Mann and M. L. Goldberger, *Physical Review Letters* 9 (1962): 275; J. D. Bjorken and T. T. Wu, *Physical Review* 130 (1963): 2566; M. Gell-Mann, M. L. Goldberger, F. E. Low, and F. Zachariasen, *Physics Letters* 4 (1963): 265; T. L. Trueman and T. Yao, *Physical Review* 132 (1963): 2741; M. Gell-Mann, M. L. Goldberger, F. E. Low, E. Marx, and F. Zachariasen, *Physical Review* 133 (1964): B145; M. Gell-Mann, M. L. Goldberger, F. E. Low, V. Singh, and F. Zachariasen, *Physical Review* 133 (1964): B161; J. C. Polkinghorne, *Journal of Mathematical Physics* 5 (1964): 431; R. J. Eden, P. V. Landshoff, D. I. Olive, and J. C. Polkinghorne, *The Analytic S-Matrix* (Cambridge: Cambridge University Press, 1966).

11 Fermion-Fermion Scattering and Tower Diagrams

11.1 Introduction

We saw in chapter 6 that in the Abelian gauge field theory uncanceled $is(\ln s)^n$ terms appear in the $4(n + 1)$th-order fermion-fermion elastic scattering amplitude. This implies that, according to the optical theorem, the total cross section for fermion-fermion scattering has $4(n + 1)$th-order terms that approach infinity as $(\ln s)^n$ as $s \to \infty$.

This result marks the first major deviation from the conclusions reached on the basis of models from s-channel potential scattering. Furthermore, it is not a peculiarity of an Abelian gauge field theory but rather has an unambiguous physical basis: the creation of pionization products. It is for this reason that we studied pionization in some detail in chapter 5.

We purposely omitted in chapter 6 the details of calculations of the uncanceled logarithms; they are given in this chapter. With the method of approximation described in the preceding chapter, it is a simple task to calculate these logarithms. Indeed, we begin systematically by examining all fourth-order and sixth-order diagrams for fermion-fermion scattering in the Abelian gauge field theory and show that there are no uncanceled logarithms up to the sixth order. Then we calculate the $s \ln s$ term in the eighth order as well as the $s \ln^n s$ term in the $4(n + 1)$th order [1,2].

11.2 Fourth-Order Diagrams in Fermion-Fermion Scattering

Some of the fourth-order diagrams for fermion-fermion scattering are illustrated in figure 11.1. The other fourth-order diagrams can be obtained from them by interchanging the two incident particles. The amplitudes from these latter diagrams are only of the order of s^0, as the exchanged vector mesons must carry large momentum transfers and their propagators are small. Therefore these amplitudes are negligible in the high-energy limit $s \to \infty$ with t fixed.

It is obvious that the first three diagrams in figure 11.1 merely modify the second-order one-vector-meson exchange amplitude by giving the first fermion the form factor (with Δ in the x direction)

$$F_1(\Delta) - i\Delta F_2(\Delta)\sigma_2, \tag{11.1}$$

where $\frac{1}{2}\sigma_2$ is the y component of the fermion spin and F_1 and F_2 are the usual form factors for the fermion.

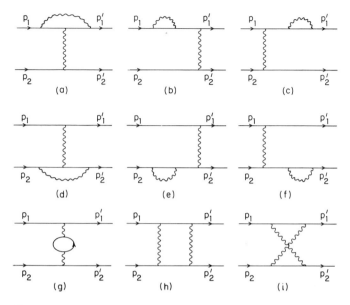

Figure 11.1
Some fourth-order diagrams for fermion-fermion elastic scattering. All other fourth-order diagrams for this reaction can be obtained from these diagrams by interchanging the two incident particles. The wavy lines represent vector mesons, and the solid lines represent fermions.

Similarly, the diagrams in figures 11.1d–f merely modify the second-order one-vector-meson exchange amplitude by giving the second fermion a form factor. Thus the sum of the one-vector-meson exchange amplitude and the amplitudes from the diagrams in figures 11.1a–f is equal to the former amplitude times the form factors of the two fermions, valid up to the fourth perturbative order. Because the form factor is a function of t only, the diagrams in figures 11.1a–f give an amplitude proportional to s without any logarithms.

Next we turn to figure 11.1h. According to our calculation in chapter 9, this box diagram gives an amplitude of the order of $s \ln s$. However, the sum of the amplitudes from the last two diagrams in figure 11.1 is proportional to s without any logarithms. This has already been shown in chapter 3. Thus the $s \ln s$ terms from figures 11.1h and 11.1i cancel. This cancellation can also be shown in the following simple way. The two diagrams are related to each other by $s \leftrightarrow u$. Thus, if one of the amplitudes asymptotically approaches $as \ln s$, where a is a constant, then the other amplitude asympto-

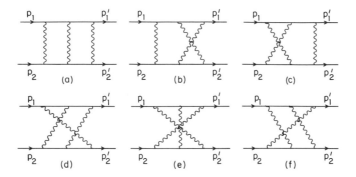

Figure 11.2
The three-vector-meson exchange diagrams for fermion-fermion elastic scattering.

tically approaches $au \ln u$. Because in the high-energy limit

$$u \simeq -s,$$

$au \ln u$ is approximately $-as \ln s$. Thus the logarithms cancel.

Finally, the diagram in figure 11.1g merely modifies the coupling constant and the vector meson propagator, which is a function of t only. Thus the corresponding amplitude is of the order of s.

11.3 Sixth-Order Diagrams in Fermion-Fermion Scattering

In this section we discuss all sixth-order diagrams of fermion-fermion scattering in the Abelian gauge field theory. We show that the sum of these sixth-order amplitudes is again of the order of s without any logarithms.

Three-Vector-Meson Exchange

A class of sixth-order diagrams for fermion-fermion scattering is illustrated in figure 11.2. They are the diagrams of three-vector-meson exchange. It was shown in the preceding chapter that each of these diagrams gives an amplitude of the order of $s \ln s$. As we showed in chapter 3, however, the sum of all three-vector-meson exchange amplitudes is of the order of s. Thus the logarithmic factors of s are canceled in the sum.

Impact Factor and Form Factor

Another class of sixth-order diagrams is illustrated in figures 11.3 and 11.4. We show that the sum of amplitudes from these diagrams is proportional

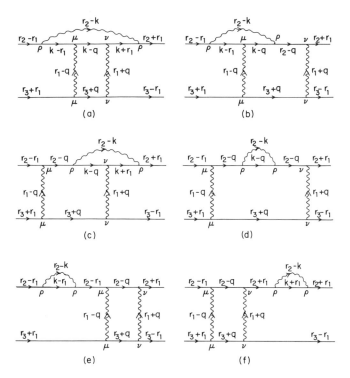

Figure 11.3
Some sixth-order diagrams for fermion-fermion elastic scattering.

to s. In fact, it is equal to the two-vector-meson exchange amplitude times the lowest-order radiative correction to the fermion form factor. There are, of course, diagrams that can be obtained from those in figures 11.3 and 11.4 by turning them upside down. The amplitude from the sum of such diagrams similarly modifies the form factor of the lower fermion. Thus the diagrams in figures 11.3 and 11.4 plus those obtained by turning them upside down merely modify the two-vector-meson exchange amplitude by the form factors of the two fermions, in exactly the same way the first six diagrams in figure 11.1 modify the one-vector-meson exchange amplitude.

Let us do the calculation. We use the notation for the external momenta:

$$p_1 \equiv r_2 - r_1, \qquad p_1' \equiv r_2 + r_1,$$
$$p_2 \equiv r_3 + r_1, \qquad p_2' \equiv r_3 - r_1. \qquad (11.2)$$

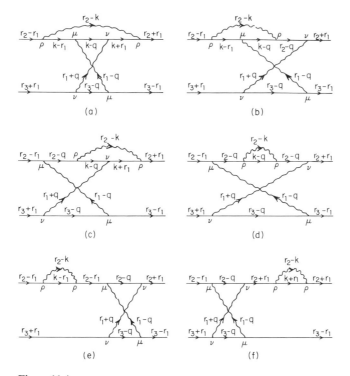

Figure 11.4
The sixth-order diagrams for fermion-fermion elastic scattering obtained from the diagrams in figure 11.3 by crossing the two exchanged vector meson lines.

(Note that r denotes momentum, not position.) Thus

$$r_1 = \tfrac{1}{2}\Delta, \qquad r_2 = \tfrac{1}{2}(p_1 + p_1'), \qquad r_3 = \tfrac{1}{2}(p_2 + p_2'). \tag{11.3}$$

This notation explicitly exhibits symmetry between the incoming and the outgoing particles:

$$p_1 \leftrightarrow p_1', \quad p_2 \leftrightarrow p_2' \Rightarrow r_1 \to -r_1, \quad \text{with } r_2, r_3 \text{ unchanged.}$$

This notation is slightly cumbersome, but in long calculations it sometimes has an advantage. Although it is not necessary, we choose to adopt this notation for the present calculation. Note that the momenta of the two exchanged vector mesons are $r_1 + q$ and $r_1 - q$, respectively, not q and $\Delta - q$ as previously denoted.

We first observe that diagrams e and f in both figure 11.3 and figure 11.4 simply contribute to an overall multiplicative constant to the lowest-order two-vector-meson exchange amplitude. Therefore let us imagine that we have carried out all renormalization procedures; then these diagrams can be ignored and suitable renormalization constants should be added to the vertex and self-energy functions.

Next we calculate the sum of amplitudes from the diagrams in figures 11.3a and 11.4a. Aside from the propagator corresponding to the lower virtual fermion, all factors in these two amplitudes are the same. Thus we can ignore the diagram in figure 11.4a if we replace the propagator of the lower virtual fermion in the diagram in figure 11.3a, together with the two γ's from the vertices, by

$$\frac{\gamma_\nu(\rlap{/}r_3 + \rlap{/}q + m)\gamma_\mu}{(r_3 + q)^2 - m^2 + i\varepsilon} + \frac{\gamma_\mu(\rlap{/}r_3 - \rlap{/}q + m)\gamma_\nu}{(r_3 - q)^2 - m^2 + i\varepsilon}$$

$$\sim \frac{2r_{3\mu}r_{3\nu}}{m}\left[\frac{1}{(r_3 + q)^2 - m^2 + i\varepsilon} + \frac{1}{(r_3 - q)^2 - m^2 + i\varepsilon}\right]$$

$$\sim \frac{-2i\pi r_{3\mu}r_{3\nu}}{m\omega}\delta(q_+). \tag{11.4}$$

The sum of the scattering amplitudes corresponding to the diagrams in figures 11.3a and 11.4a is therefore asymptotically equal to

$$\frac{ig^6}{2(2\pi)^7 m\omega}\int d^2q_\perp\, dq_-\, d^4k$$

$$\times \frac{\gamma_\rho(\rlap{/}k + \rlap{/}r_1 + m)\gamma_3(\rlap{/}k - \rlap{/}q + m)\gamma_3(\rlap{/}k - \rlap{/}r_1 + m)\gamma_\rho}{[(r_2 - k)^2 - \lambda^2][(k + r_1)^2 - m^2][(k - q)^2 - m^2][(k - r_1)^2 - m^2]}$$

$$\times [(\mathbf{r}_1 + \mathbf{q}_\perp)^2 + \lambda^2]^{-1}[(\mathbf{r}_1 - \mathbf{q}_\perp)^2 + \lambda^2]^{-1}, \tag{11.5}$$

where the gamma matrices are to be inserted between the spinors of the upper fermion and where q_+ is understood to be zero in the integrand of expression (11.5). Let us cast expression (11.5) into the impact-factor representation (9.1). Then with $\mathscr{I}^b$ identified with $g^2/2m$, the impact factor of the upper fermion is equal to

$$\mathscr{I}_a = \frac{ig^2}{\omega}\int_{-\infty}^{\infty}\frac{dq_-}{2\pi}(-ig^2)\int\frac{d^4k}{(2\pi)^4}$$

$$\times \frac{\gamma_\rho(\rlap{/}k + \rlap{/}r_1 + m)\gamma_0(\rlap{/}k - \rlap{/}q + m)\gamma_0(\rlap{/}k - \rlap{/}r_1 + m)\gamma_\rho}{[(r_2 - k)^2 - \lambda^2][(k + r_1)^2 - m^2][(k - q)^2 - m^2][(k - r_1)^2 - m^2]}\Bigg|_{q_+ = 0}, \tag{11.6}$$

where we have made the approximation $\slashed{r}_3 \sim \omega(\gamma_0 + \gamma_3) \sim 2\omega\gamma_0$. The right-hand side of equation (11.6) is the contribution from the diagrams in figures 11.3a and 11.4a to the impact factor of the upper fermion.

Similarly, the contribution from the diagrams in figures 11.3b and 11.4b to the impact factor of the upper fermion is

$$\mathscr{I}_b = \frac{ig^2}{\omega} \int_{-\infty}^{\infty} \frac{dq_-}{2\pi} [(r_2 - q)^2 - m^2]^{-1} \gamma_0 (\slashed{r}_2 - \slashed{q} + m)$$

$$\times \left\{ -ig^2 \int \frac{d^4k}{(2\pi)^4} \frac{\gamma_\rho (\slashed{k} - \slashed{q} + m)\gamma_0 (\slashed{k} - \slashed{r}_1 + m)\gamma_\rho}{[(r_2 - k)^2 - \lambda^2][(k - q)^2 - m^2][(k - r_1)^2 - m^2]} - L\gamma_0 \right\} \Bigg|_{q_+ = 0} .$$
(11.7)

The contribution from the diagrams in figures 11.3c and 11.4c to the impact factor of the upper fermion is

$$\mathscr{I}_c = \frac{ig^2}{\omega} \int_{-\infty}^{\infty} \frac{dq_-}{2\pi} [(r_2 - q)^2 - m^2]^{-1}$$

$$\times \left\{ -ig^2 \int \frac{d^4k}{(2\pi)^4} \frac{\gamma_\rho (\slashed{k} + \slashed{r}_1 + m)\gamma_0 (\slashed{k} - \slashed{q} + m)\gamma_\rho}{[(r_2 - k)^2 - \lambda^2][(k + r_1)^2 - m^2][(k - q)^2 - m^2]} - L\gamma_0 \right\}$$

$$\times (\slashed{r}_2 - \slashed{q} + m)\gamma_0 \Bigg|_{q_+ = 0} . \quad (11.8)$$

And the contribution from the diagrams in figures 11.3d and 11.4d to the impact factor of the upper fermion is

$$\mathscr{I}_d = \frac{ig^2}{\omega} \int_{-\infty}^{\infty} \frac{dq_-}{2\pi} \frac{\gamma_0 (\slashed{r}_2 - \slashed{q} + m)}{(r_2 - q)^2 - m^2}$$

$$\times \left\{ -ig^2 \int \frac{d^4k}{(2\pi)^4} \frac{\gamma_\rho (\slashed{k} - \slashed{q} + m)\gamma_\rho}{[(r_2 - k)^2 - \lambda^2][(k - q)^2 - m^2]} - A + L(\slashed{r}_2 - \slashed{q} - m) \right\}$$

$$\times \frac{(\slashed{r}_2 - \slashed{q} + m)\gamma_0}{(r_2 - q)^2 - m^2} \Bigg|_{q_+ = 0} . \quad (11.9)$$

In equations (11.7) through (11.9), A and L are the usual renormalization constants for the lowest-order radiative corrections in Abelian gauge theory. For example, $L\gamma_0$ is equal to the (divergent) integral inside the braces of equation (11.7) at the point $q = r_1$. Thus all the expressions in braces in equations (11.7) through (11.9) are finite. The fermion impact factor is therefore given by

$$\mathscr{I}^f = \frac{g^2}{2m} + \lim_{\omega \to \infty} (\mathscr{I}_a + \mathscr{I}_b + \mathscr{I}_c + \mathscr{I}_d), \tag{11.10}$$

up to the fourth order, provided that this limit exists.

Instead of studying the limit in equation (11.10) directly, we use the following procedure. We introduce a cutoff in the integration over k so that both A and L are finite. With such a finite cutoff, the right-hand side of equation (11.10) is guaranteed to be finite. The question is simply whether it approaches a limit as the cutoff approaches infinity.

We begin by using in equation (11.6) the approximation

$$\gamma_0(\not{k} - \not{q} + m)\gamma_0 \sim 2k_0\gamma_0. \tag{11.11}$$

Also, by symmetrizing q and $-q$, we have

$$\int_{-\infty}^{\infty} \frac{dq_-}{(k-q)^2 - m^2 + i\varepsilon}$$

$$\to \frac{1}{2} \int_{-\infty}^{\infty} dq_- \left[\frac{1}{(k-q)^2 - m^2 + i\varepsilon} + \frac{1}{(k+q)^2 - m^2 + i\varepsilon} \right]$$

$$= -\frac{i\pi}{|k_+|} = -\frac{i\pi}{k_+}. \tag{11.12}$$

The last step of equation (11.12) is justified, as the integral over k_- in the following relation (11.13) vanishes if k_+ is negative. With relations (11.11) and (11.12), we reduce equation (11.6) to

$$\mathscr{I}_a \sim \frac{g^2}{2\omega} \left\{ -ig^2 \int \frac{d^4k}{(2\pi)^4} \right.$$

$$\times \left. \frac{\gamma_\rho(\not{k} + \not{r}_1 + m)\gamma_0(\not{k} - \not{r}_1 + m)\gamma_\rho}{[(r_2 - k)^2 - \lambda^2 + i\varepsilon][(k+r_1)^2 - m^2 + i\varepsilon][(k-r_1)^2 - m^2 + i\varepsilon]} \right\}. \tag{11.13}$$

The quantity inside the braces is recognized as the unrenormalized vertex function. Thus

$$\mathscr{I}_a \sim \frac{g^2}{2\omega} [\Lambda_0(r_2 + r_1, r_2 - r_1) + L\gamma_0], \tag{11.14}$$

where Λ_0 is the lowest-order renormalized radiative correction to the fermion vertex function.

Our attention is turned next to $\mathscr{I}_b$ as given by equation (11.7). Consider first the first term inside the braces. Because we can neglect the $\not{q}$ in the numerator, the integration over q_- involves the integral

$$\int_{-\infty}^{\infty} \frac{dq_-}{[(r_2 - q)^2 - m^2 + i\varepsilon][(k - q)^2 - m^2 + i\varepsilon]}\bigg|_{q_+ = 0}$$

$$\sim \int_{-\infty}^{\infty} \frac{dq_-}{[2\omega q_- + r_2^2 - \mathbf{q}_\perp^2 - m^2 + i\varepsilon][-k_+ q_- + k_+ k_- - (\mathbf{k}_\perp - \mathbf{q}_\perp)^2 - m^2 + i\varepsilon]}$$

$$= 0. \tag{11.15}$$

The last step in relation (11.15) is obtained because k_+ is positive and because we can close the contour of integration in the lower half-plane of q_-, where the integrand is analytic. Thus the term $L\gamma_0$ inside the braces of equation (11.7) is the only one that survives after the integration over q_-. By using relations (11.11) and (11.12) with k replaced by r_2, we then obtain

$$\mathscr{I}_b \sim \frac{g^2}{2\omega} L\gamma_0. \tag{11.16}$$

Similarly,

$$\mathscr{I}_c \sim \frac{g^2}{2\omega} L\gamma_0, \tag{11.17}$$

$$\mathscr{I}_d \sim \frac{g^2}{2\omega} L\gamma_0. \tag{11.18}$$

The sum of relation (11.14) and relations (11.16) through (11.18) gives

$$\mathscr{I}_a + \mathscr{I}_b + \mathscr{I}_c + \mathscr{I}_d \sim \frac{g^2}{2\omega} \Lambda_0(r_2 + r_1, \ r_2 - r_1). \tag{11.19}$$

The cancellation of L shows that the radiative correction to the impact factor is defined as the cutoff approaches infinity. From relations (11.10) and (11.19), we get

$$\mathscr{I}^f = \lim_{\omega \to \infty} \frac{g^2}{2\omega} \bar{u}(\mathbf{r}_2 + \mathbf{r}_1)[\gamma_0 + \Lambda_0(r_2 + r_1, \ r_2 - r_1)]u(\mathbf{r}_2 - \mathbf{r}_1)$$

$$= \lim_{\omega \to \infty} \frac{g^2}{2\omega} \bar{u}(\mathbf{r}_2 + \mathbf{r}_1)\Gamma_0(r_2 + r_1, \ r_2 - r_1)u(\mathbf{r}_2 - \mathbf{r}_1), \tag{11.20}$$

up to the fourth order in g, where $\Gamma_\mu(r_2 + r_1, \ r_2 - r_1)$ is the renormalized vertex function of the fermion. Thus, up to the fourth order of g, the fermion impact factor is proportional to the fermion vertex function [3].

Other Sixth-Order Diagrams

Another class of sixth-order diagrams is illustrated in figures 11.5, 11.6, and 11.7. They are diagrams of one-vector-meson exchange. In addition, there are also the diagrams obtained by turning figures 11.5 and 11.6 upside down. They simply modify the vertex functions and the vector meson propagator in the one-vector-meson exchange amplitude. Because the vertex functions and the propagator depend on t but not on s, the amplitudes corresponding to these diagrams are of the order of s. The only other sixth-order diagrams left are the ones illustrated in figure 11.8. These diagrams simply modify the vector meson propagators in the two-vector-meson exchange amplitude, in addition to contributing to charge renormalization. This means that the sum of amplitudes from these diagrams is

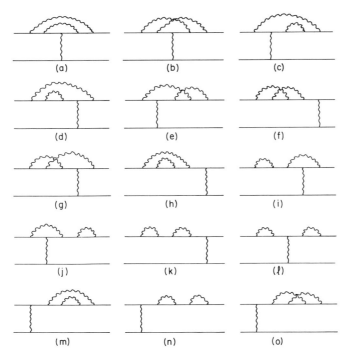

Figure 11.5
Some sixth-order diagrams with the exchange of one vector meson between the two fermions.

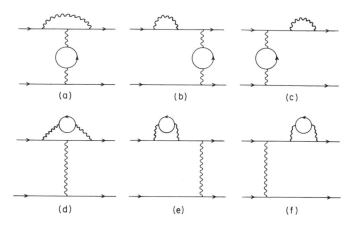

Figure 11.6
More sixth-order diagrams with the exchange of one vector meson between the two fermions.

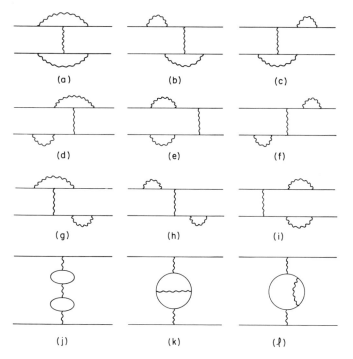

Figure 11.7
More sixth-order diagrams with the exchange of one vector meson between the two fermions.

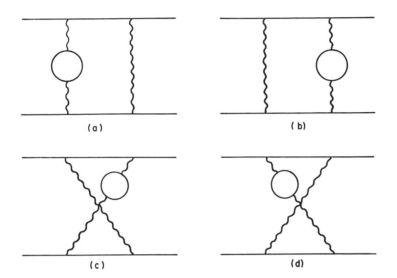

Figure 11.8
Some sixth-order diagrams in which a bare vector meson and a dressed vector meson are exchanged between the two fermions.

of the impact-factor representation (9.1), with $\mathscr{I}^a$ and $\mathscr{I}^b$ both given by the lowest-order fermion impact factor and with the vector-meson propagators $[(\mathbf{r}_1 + \mathbf{q}_\perp)^2 + \lambda^2]^{-1}$ or $[(\mathbf{r}_1 - \mathbf{q}_\perp)^2 + \lambda^2]^{-1}$ replaced by the lowest-order radiative correction to the vector meson propagators. Because the resulting integral over $\mathbf{q}_\perp$ is convergent, this amplitude is again of the order of s.

In conclusion, we have shown that none of the sixth-order diagrams for fermion-fermion scattering gives a scattering amplitude that grows faster than s as $s \to \infty$ with fixed t. Furthermore, if we consider only diagrams for fermion-fermion scattering up to the sixth order, adding radiative corrections to both the one-vector-meson exchange amplitude and the two-vector-meson exchange amplitude has simply the following consequences: (1) It gives an overall factor (equal to the vertex function in *both* cases) to the corresponding scattering amplitudes; (2) it modifies the bare vector meson propagator into the dressed vector meson propagator. Result 1 can be generalized. The incident fermion can be considered a bare fermion plus a cloud of pairs and vector mesons. For the diagrams in which the bare fermion is the only one interacting with the other incident particle, result 1 remains true. (For further discussions on this point, see pp. 159–161.)

11.4   Eighth-Order Diagrams in Fermion-Fermion Scattering

In the preceding section we gave a systematic discussion of all sixth-order diagrams for fermion-fermion scattering. In this section we turn our attention to the eighth-order diagrams. There are almost one thousand diagrams for fermion-fermion scattering of the eighth order. We will see, however, that only *three* of the diagrams are interesting.

In order to give a systematic discussion, let us classify the eighth-order diagrams by the number of vector mesons exchanged.

Four Exchanged Vector Mesons

In the eighth order there are twenty-four Feynman diagrams with four vector mesons exchanged. All these vector mesons are "bare," that is, there are no self-energy parts on the virtual vector meson lines. We showed in chapter 3 that the sum of such four-photon exchange amplitudes is of the order of s.

Three Exchanged Vector Mesons

There are 120 diagrams in which three bare vector mesons are exchanged between the two incident fermions. They are obtained from the three-vector-meson exchange diagrams shown in figure 11.2 by adding a radiative correction to one of the fermion lines. From a discussion similar to the one given in the preceding section, we find that the sum of the eighth-order diagrams of three-vector-meson exchange merely adds a vertex factor to the lowest-order three-vector-meson exchange amplitude.

In addition, there are eighteen Feynman diagrams in which the two incident fermions exchange two bare vector mesons and one "dressed" vector meson that has a lowest-order self-energy part. One such diagram is illustrated in figure 11.9. These diagrams merely contribute to the re-

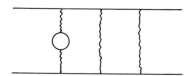

Figure 11.9
An eighth-order diagram in which two bare vector mesons and a dressed vector meson are exchanged between the two fermions.

normalization of the coupling constant and the dressing of the exchanged vector meson but do not change the dependence on *s*.

Finally, there are six diagrams in which a fermion-antifermion pair is created, and this pair exchanges three vector mesons with one of the incident fermions and one vector meson with the other incident fermion. Three such diagrams are illustrated in figure 11.10, and the other three can be obtained by turning the diagrams in figure 11.10 upside down. The scattering amplitude corresponding to these diagrams is proportional to *s*. This is because in each of these diagrams, one of the incident fermions exchanges only one vector meson with the created pair. Thus the rest of the diagram can be viewed as the absorption of this vector meson by the other incident fermion by means of a fermion loop—a radiative correction to the vertex of the other fermion. The scattering amplitude is therefore equal to the second-order fermion-fermion scattering amplitude times a vertex function depending only on *t* and is hence of the order of *s*.

One Exchanged Vector Meson

The class of diagrams with two exchanged vector mesons contains some interesting diagrams, but we delay the discussion of them until pp. 159–161. Here we discuss the diagrams with one exchanged vector meson. These are the kind of diagram that can be separated into two disconnected parts by cutting one virtual vector meson line. Indeed, the diagrams in figure 11.10 are also of this kind.

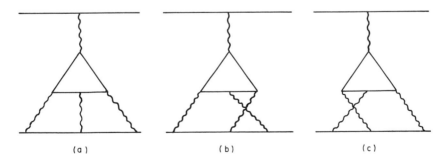

(a)   (b)   (c)

Figure 11.10
Some eighth-order diagrams with a fermion loop. This loop is joined to the upper (resp. lower) fermion by one (resp. three) vector meson(s).

From the discussion in the preceding subsection, the scattering amplitude for a diagram of this kind is equal to the second-order fermion-fermion scattering amplitude times two vertex functions corresponding to the two disconnected parts of the diagram obtained by cutting the virtual vector meson line. Thus the scattering amplitude for a diagram of this kind is of the order of s.

We enumerate these diagrams in what follows. In addition to the six diagrams already discussed in the preceding subsection, we have:

1. One hundred and five diagrams in which a bare vector meson is exchanged and three virtual bare vector mesons are emitted and reabsorbed by the upper fermion.

2. One hundred and five diagrams in which a bare vector meson is exchanged and three virtual bare vector mesons are emitted and reabsorbed by the lower fermion.

3. Ninety diagrams in which a bare vector meson is exchanged and emission and reabsorption of virtual bare vector mesons occur for both the upper and the lower fermion.

4. Thirty-nine diagrams in which a bare vector meson is exchanged and either one dressed vector meson or one dressed vector meson plus one bare vector meson is emitted and reabsorbed by the upper fermion.

5. Thirty-nine diagrams in which a bare vector meson is exchanged and either one dressed vector meson or one dressed vector meson plus one bare vector meson is emitted and reabsorbed by the lower fermion.

6. Eighteen diagrams in which a bare vector meson is exchanged and one bare vector meson is emitted and reabsorbed by one of the fermions; also one dressed vector meson is emitted and reabsorbed by the other fermion.

7. Forty-five diagrams in which a dressed vector meson with lowest-order (second-order) radiative correction is exchanged.

8. Eighteen diagrams in which a dressed vector meson with fourth-order radiative correction is exchanged.

9. Fourteen diagrams in which a dressed vector meson with sixth-order radiative correction is exchanged.

Altogether there are 473 diagrams in these nine categories.

Two Exchanged Vector Mesons

We list the diagrams with two exchanged vector mesons:

1. Ninety diagrams in which two bare vector mesons are exchanged and two bare vector mesons are emitted and reabsorbed by the upper fermion.

2. Ninety diagrams in which two bare vector mesons are exchanged and two bare vector mesons are emitted and reabsorbed by the lower fermion.

3. Twelve diagrams in which two bare vector mesons are exchanged and one dressed vector meson is emitted and reabsorbed by the upper fermion.

4. Twelve diagrams in which two bare vector mesons are exchanged and one dressed vector meson is emitted and reabsorbed by the lower fermion.

5. Seventy-two diagrams in which two bare vector mesons are exchanged and both fermions emit and reabsorb virtual vector mesons.

6. Forty-eight diagrams in which one bare vector meson and one dressed vector meson with second-order radiative correction are exchanged.

7. Twelve diagrams in which one bare vector meson and one dressed vector meson with fourth-order radiative correction are exchanged.

8. Two diagrams in which two dressed vector mesons are exchanged.

9. Three diagrams in which the two exchanged vector mesons interact by creating a fermion-antifermion pair (see figure 11.12).

There are 341 diagrams in total.

Of these 341 diagrams, two types are interesting. The first type consists of twelve diagrams: Six of them are shown in figure 11.11, whereas the other six can be obtained by turning the diagrams in figure 11.11 upside down. The second type consists of the three diagrams illustrated in figure 11.12.

The Feynman diagrams in figure 11.11 give a fermion impact factor that depends not only on r_1 but also on $q_\perp$ [3]. In other words, the impact factor is no longer proportional to the form factor when fourth-order corrections are considered. To understand the reason for this nonproportionality, we redraw figure 11.11a in figure 11.13. We see that figure 11.13 can be viewed as a process in which both a fermion (line 2) *and* an antifermion (line 1) exchange vector mesons with the lower fermion. This diagram is thus intrinsically different from a vertex diagram, in which there is only one vertex (and hence only one particle) interacting with an external field. The

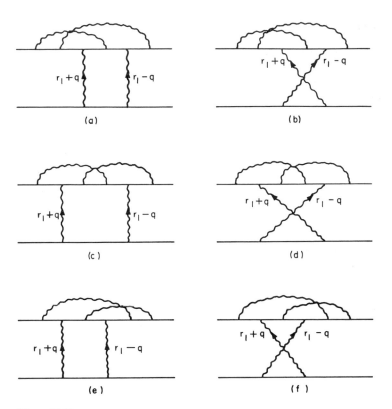

Figure 11.11
The eighth-order diagrams that give to the impact factor of the upper fermion terms
dependent on $\mathbf{q}_\perp$ as well as on $\mathbf{r}_1$.

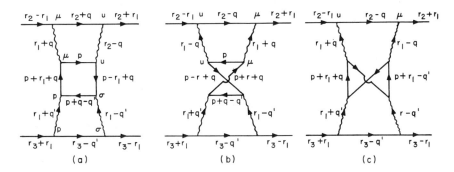

Figure 11.12
The diagrams that give the dominant amplitude, of the order of $s \ln s$, to the eighth-order
amplitude of fermion-fermion scattering. Note that there is a fermion loop in each diagram.

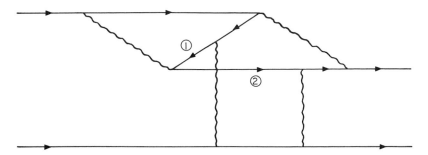

Figure 11.13
Figure 11.11a redrawn to show that the upper incident fermion can turn into a virtual state
of a fermion and a fermion-antifermion pair.

scattering amplitude corresponding to this diagram is still of the impact-
factor representation (9.1) and is of the order of s.

The three diagrams illustrated in figure 11.12 are most interesting and
are discussed in detail in the next section.

11.5   Pair Production and Logarithmic Factor of the Energy

In this section we turn our attention to the three diagrams in figure 11.12.
We show that the *sum of amplitudes corresponding to these diagrams is of
the order of* $s \ln s$ [2, 4].

It may be worthwhile to point out that an individual Feynman diagram
in figure 11.12 gives an amplitude of the order of s^2. Furthermore, the sum
of s^2 terms from the three diagrams does not vanish. This s^2 term is related
to gauge invariance; when we use Feynman rules to write the scattering
amplitude corresponding to these three diagrams, the result is not gauge
invariant. This is because the fourth-order V-V scattering amplitude as
given by the Feynman rules does not vanish when we set the momenta of
all four external vector mesons to 0. To understand this peculiarity, let us
review the usual arguments about the gauge invariance of the sum of
contributions from Feynman diagrams. Consider a set of N Feynman
diagrams that differ from each other only in the position at which one
particular vector meson line is attached to a fermion line. We replace the
polarization vector of this particular vector meson by its momentum. The
resulting amplitude corresponding to any one of these diagrams can be
split into two terms, each of which has one fermion propagator canceled.

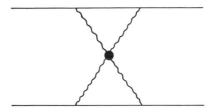

Figure 11.14
The diagram representing a subtraction term for the three diagrams in figure 11.12 and preserving gauge invariance.

After a shift in momenta these $2N$ terms cancel pairwise, and we thus establish gauge invariance. This procedure fails, however, in the case of the fourth-order V-V scattering amplitude. This is because each of the $2N$ terms is linearly divergent, and shifting a linearly divergent integral gives an extra, nonzero term. In order to restore gauge invariance, this extra, undesirable term must be eliminated. To accomplish this, we simply add the diagram of figure 11.14 to the three diagrams in figure 11.12, where the dot in figure 11.14 gives a factor equal to the negative of the fourth-order V-V scattering amplitude given by the Feynman rules at the point where the momenta of all four external photons are equal to 0. (This is the only known case in which a straightforward application of the Feynman rules gives an incorrect answer. This problem can also be avoided by dimensional regularization.) The sum of the s^2 terms from the four diagrams in figures 11.12 and 11.14 is equal to 0.

The contribution from the diagram in figure 11.14 is especially easy to take care of in the method of momentum integration outlined in the preceding chapter. We note that a fermion propagator, considered as a function of its plus momentum, does not vanish as the plus momentum goes to infinity:

$$\frac{1}{\not{p} - m} = \frac{\not{p} + m}{p^2 - m^2} = \frac{\not{p} - m}{p_+ p_- - \mathbf{p}_\perp^2 - m^2} \xrightarrow[p_+ \to \infty]{} \text{constant.} \tag{11.21}$$

It follows that, as we close the contour of the p_+-integration in figure 11.12a, for example, there is a contribution at $p_+ = \infty$. This contribution is canceled, however, by the corresponding term in figure 11.14. Therefore, when we carry out the integration over p_+, we can ignore the contribution from figure 11.14, provided that we also ignore the contribution of $p_+ = \infty$.

With this understanding let us calculate the amplitudes for figure 11.12a.

This diagram has only one flow diagram, which is illustrated in figure 11.15a. The integration over $\prod_{i=1}^{3} dq_{i+}$ is carried out by enclosing the poles as indicated in the figure. We have (see equation (10.5))

$$q_{1+} = \frac{a_9}{Q_1}, \qquad q_{2+} = \frac{a_8}{Q_2}, \qquad q_{3+} \simeq -\frac{a_9}{Q_1} - \frac{a_8}{Q_2},$$

where $Q_i \equiv q_{i-}$, $i = 1, 2, 3$. The D function can easily be written:

$$D_a \simeq \left[-Q_3 \left(\frac{a_9}{Q_1} + \frac{a_8}{Q_2} \right) - a_1 \right] \left[-Q_3 \left(\frac{a_9}{Q_1} + \frac{a_8}{Q_2} \right) - a_2 \right]$$

$$\times \left[-(Q_2 + Q_3) \frac{a_9}{Q_1} - a_3 \right] \left[-(Q_2 + Q_3) \frac{a_9}{Q_1} - a_4 \right]$$

$$\times (-a_5)(-a_6)(2\omega Q_3 + i\varepsilon)(Q_2)(Q_1)(2\omega), \tag{11.22}$$

where we have made the approximation of $Q_i \ll \omega$, $i = 1, 2, 3$. We note from the directions of the arrows on lines 7, 8, and 9 that Q_1, Q_2, Q_3 are all positive. The numerator of the Feynman integral for figure 11.12a is

$$N_a \simeq \left(\frac{2r_{2\mu}r_{2\nu}}{m} \right) \left(\frac{2r_{3\rho}r_{3\sigma}}{m} \right)$$

$$\times (-1)(2) \operatorname{Tr} [\gamma^\mu (\not{q}_2 + m) \gamma^\nu (\not{q}_2 + \not{q}_3 - \not{r}_1 + m) \gamma^\sigma (-\not{q}_1 + m)$$

$$\times \gamma^\rho (\not{r}_1 + \not{q}_2 + \not{q}_3 + m)], \tag{11.23}$$

where the first (resp. second) factor of N_a comes from the upper (resp. lower) fermion line, the factor of 2 takes care of the two directions of a fermion loop, and the minus sign comes from the Feynman rules for a closed fermion loop. Relation (11.23) can be rewritten as

$$N_a \simeq -8 \frac{\omega^4}{m^2} \operatorname{Tr} [\gamma_-(\not{q}_2 + m) \gamma_-(\not{q}_2 + \not{q}_3 - \not{r}_1 + m) \gamma_+(-\not{q}_1 + m)$$

$$\times \gamma_+(\not{r}_1 + \not{q}_2 + \not{q}_3 + m)]$$

$$= -8 \frac{\omega^4}{m^2} (2Q_2)(-2q_{1+}) \operatorname{Tr} [\gamma_-(\not{q}_2 + \not{q}_3 - \not{r}_1 + m) \gamma_+(\not{r}_1 + \not{q}_2 + \not{q}_3 + m)]$$

$$= \frac{2s^2}{m^2} Q_2 \frac{a_9}{Q_1} \operatorname{Tr} [\gamma_-(\not{q}_{2\perp} + \not{q}_{3\perp} - \not{r}_1 + m) \gamma_+(\not{r}_1 + \not{q}_{2\perp} + \not{q}_{3\perp} + m)].$$

$$\tag{11.24}$$

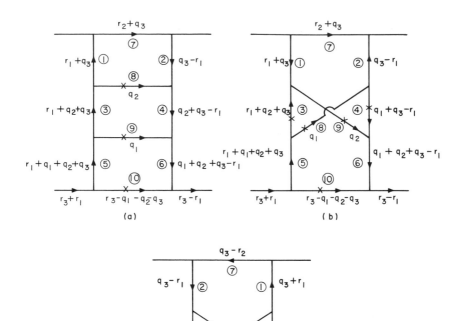

Figure 11.15
(a) The flow diagram corresponding to figure 11.12a. (b, c) The two flow diagrams
corresponding to figure 11.12c.

In equation (11.24) we have made use of the facts

$$\gamma_+^2 = \gamma_-^2 = 0,$$

$$\gamma_+ \gamma_\perp + \gamma_\perp \gamma_+ = \gamma_- \gamma_\perp + \gamma_\perp \gamma_- = 0.$$

Also,

$$\not{q}_\perp \equiv -\mathbf{q}_\perp \cdot \boldsymbol{\gamma}_\perp.$$

Figure 11.15a is a ladder diagram. As we saw in the preceding chapter, this diagram gives a $\ln^3 s$ term if the numerator is set to unity, with the region giving the dominant contribution being $\omega \gg Q_1 \gg Q_2 \gg Q_3$. However, N_a in equation (11.24) is proportional to Q_2/Q_1, which is small if $Q_1 \gg Q_2$. Thus the contributing region for the Q_1 diagram in figure 11.12a is

$$\omega \gg Q_1 \gg Q_3, \qquad Q_2/Q_1 = O(1). \tag{11.25}$$

From relations (11.25) and the formulas preceding equation (11.22), we find that q_{1+}, q_{2+}, and q_{3+} are comparable. Referring to figure 11.12a, we see that both the plus and the minus momenta of the created fermion and the created antifermion in the fermion loop are comparable. Furthermore, the plus momenta of these created particles are much less than that of the upper fermion, and the minus momenta of these particles are much less than that of the lower fermion.

Let us denote

$$Q_1 \equiv xQ, \qquad Q_2 \equiv (1 - x)Q, \tag{11.26}$$

where $0 < x < 1$. Then

$$dQ_1\, dQ_2 = Q\, dQ\, dx.$$

Also,

$$N_a \simeq 2 \frac{s^2}{m^2} \frac{1-x}{x} a_9 \operatorname{Tr}[\gamma_-(\not{q}_{2\perp} + \not{q}_{3\perp} - \not{r}_1 + m)\gamma_+(\not{r}_1 + \not{q}_{2\perp} + \not{q}_{3\perp} + m)], \tag{11.27}$$

$$D_a \simeq sa_1 a_2[(1 - x)a_9 + xa_3][(1 - x)a_9 + xa_4]$$

$$\times (a_5 a_6)Q^2(1 - x)x^{-1}(Q_3 + i\varepsilon). \tag{11.28}$$

The integration over the minus momenta is easily carried out. But before we do that, let us observe that figure 11.12b differs from figure 11.12a only in the upper fermion line. Thus it has the same N, but its D function can be obtained from relation (11.28) by the replacement

$$(Q_3 + i\varepsilon) \to -Q_3.$$

Therefore the sum of the amplitudes of figures 11.12a and 11.12b is equal to the amplitude of figure 11.12a with the replacement

$$(Q_3 + i\varepsilon)^{-1} \to (Q_3 + i\varepsilon)^{-1} + (-Q_3)^{-1} \simeq -i\pi\delta(Q_3). \tag{11.29}$$

(Recall that, as discussed in chapter 10, $(-Q_3)^{-1}$ has no imaginary part, whereas $(Q_3 + i\varepsilon)^{-1}$ has the imaginary part $-i\pi\delta(Q_3)$.) We therefore get

$$\mathcal{M}_a + \mathcal{M}_b \simeq (ig^8)(-\tfrac{1}{2}i)^3 \int \prod_{i=1}^{3} \frac{d^2 q_{i\perp}}{(2\pi)^2} \int_0^1 \frac{dx}{2\pi} \int Q \frac{dQ \, dQ_3}{2\pi \, 2\pi}$$

$$\times \frac{2(s/m^2) a_9 \operatorname{Tr}[\gamma_-(\slashed{q}_{2\perp} + \slashed{q}_{3\perp} - \slashed{r}_1 + m)\gamma_+(\slashed{r}_1 + \slashed{q}_{2\perp} + \slashed{q}_{3\perp} + m)]}{a_1 a_2 a_5 a_6 [(1-x)a_9 + x a_3][(1-x)a_9 + x a_4]}$$

$$\times \frac{1}{Q^2}[-i\pi\delta(Q_3)], \quad (11.30)$$

where the first factor in this expression comes from

$$(-i)(-i)^v (i)^f (-ig)^n = ig^8$$

where $v = 4$ is the number of virtual vector meson lines, $f = 6$ is the number of virtual fermion lines, and $n = 8$ is the order of perturbation. Also, the factor $(-\tfrac{1}{2}i)^3$ in relation (11.30) comes from three integrations over $\prod_1^3 dq_{i+}$. Carrying the calculation over Q and Q_3, we get

$$\mathcal{M}_a + \mathcal{M}_b \simeq ig^8 \frac{s}{8m^2} \frac{\ln s}{(2\pi)^2} \int \prod_{i=1}^{3} \frac{d^2 q_{i\perp}}{(2\pi)^2} \int_0^1 dx$$

$$\times \frac{a_9 \operatorname{Tr}[\gamma_-(\slashed{q}_{2\perp} + \slashed{q}_{3\perp} - \slashed{r}_1 + m)\gamma_+(\slashed{r}_1 + \slashed{q}_{2\perp} + \slashed{q}_{3\perp} + m)]}{a_1 a_2 a_5 a_6 [(1-x)a_9 + x a_3][(1-x)a_9 - x a_4]}.$$

$$\tag{11.31}$$

Next we calculate the diagram in figure 11.12c. It has two flow diagrams, illustrated in figures 11.15b and 11.15c. Consider first the flow diagram in figure 11.15b. With the poles enclosed as in the figure, we have

$$q_{1+} = \frac{a_8}{Q_1}, \qquad q_{2+} = \frac{a_9}{Q_2}, \qquad q_{3+} \simeq \frac{a_8}{Q_1} - \frac{a_9}{Q_2}. \tag{11.32}$$

The D function is easily written as

$$D_c \simeq \left[-Q_3\left(\frac{a_8}{Q_1} + \frac{a_9}{Q_2}\right) - a_1 \right]\left[-Q_3\left(\frac{a_8}{Q_1} + \frac{a_9}{Q_2}\right) - a_2 \right]$$

$$\times \left[-(Q_2 + Q_3)\frac{a_8}{Q_1} - a_3 \right]\left[-(Q_1 + Q_3)\frac{a_9}{Q_2} - a_4 \right]$$

$$\times (-a_5)(-a_6)(2\omega Q_3 + i\varepsilon)(Q_1)(Q_2)(2\omega). \tag{11.33}$$

And the numerator for the diagram in figure 11.15b is

$$N_c \simeq \left(\frac{2r_{2\mu}r_{2\nu}}{m}\right)\left(\frac{2r_{3\rho}r_{3\sigma}}{m}\right)(-2)\,\mathrm{Tr}[\gamma^\nu(\not q_2 + m)\gamma^\sigma(-\not q_1 - \not q_3 + \not r_1 + m)$$

$$\times \gamma^\mu(-\not q_1 + m)\gamma^\rho(\not r_1 + \not q_2 + \not q_3 + m)]$$

$$\simeq -\frac{8\omega^4}{m^2}\,\mathrm{Tr}[\gamma_-(\not q_{2\perp} + m)\gamma_+(-\not q_{1\perp} - \not q_{3\perp} + \not r_1 + m)\gamma_-(-\not q_{1\perp} + m)$$

$$\times \gamma_+(\not r_1 + \not q_{2\perp} + \not q_{3\perp} + m)]. \tag{11.34}$$

The flow diagram in figure 11.15c gives the same numerator N_c and a D function that is obtained from D_c by replacing $(2\omega Q_3 + i\varepsilon)$ with $(-2\omega Q_3)$. Thus the sum of amplitudes from figures 11.15b and 11.15c is equal to

$$\mathcal{M}_c \simeq -\frac{isg^8 \ln s}{32m^2(2\pi)^2}\int \prod_{i=1}^3 \frac{d^2 q_{i\perp}}{(2\pi)^2}$$

$$\times \int_0^1 dx \frac{\mathrm{Tr}[\gamma_-(\not q_{2\perp}+m)\gamma_+(-\not q_{1\perp}-\not q_{3\perp}+\not r_1+m)\gamma_-(-\not q_{1\perp}+m)\gamma_+(\not r_1+\not q_{2\perp}+\not q_{3\perp}+m)]}{a_1 a_2[(1-x)a_8 + xa_3][(1-x)a_4 + xa_9]a_5 a_6}. \tag{11.35}$$

The traces in relations (11.31) and (11.35) can be further reduced. We have

$$\mathrm{Tr}[\gamma_-(\not A_\perp + m)\gamma_+(\not B_\perp + m)] = 2\,\mathrm{Tr}[(-\not A_\perp + m)(\not B_\perp + m)]$$

and

$$\mathrm{Tr}[\gamma_-(\not A_\perp + m)\gamma_+(\not B_\perp + m)\gamma_-(\not C_\perp + m)\gamma_+(\not D_\perp + m)]$$

$$= 8\,\mathrm{Tr}[(-\not A_\perp + m)(\not B_\perp + m)(-\not C_\perp + m)(\not D_\perp + m)].$$

Thus, by adding relations (11.31) and (11.35), we find that the sum of the amplitudes from the diagrams in figure 11.12 is

$$\mathcal{M}_a + \mathcal{M}_b + \mathcal{M}_c$$

$$\simeq is\ln s \int \prod_{i=1}^{2} \left[\frac{d^2 q_{i\perp}}{(2\pi)^2} \frac{1}{(\mathbf{r}_1 + \mathbf{q}_{i\perp})^2 + \lambda^2} \frac{1}{(\mathbf{r}_1 - \mathbf{q}_{i\perp})^2 + \lambda^2} \right]$$

$$\times \mathcal{I}^f K(\mathbf{r}_1, \mathbf{q}_{1\perp}, \mathbf{q}_{2\perp}) \mathcal{I}^f, \quad (11.36)$$

where

$$K(\mathbf{r}_1, \mathbf{q}_{1\perp}, \mathbf{q}_{2\perp}) = g^4 (2\pi)^{-4} \int d^2 p_\perp \int_0^1 \frac{dx}{x(\mathbf{p}_\perp + \mathbf{r}_1 + \mathbf{q}_{1\perp})^2 + (1-x)\mathbf{p}_\perp^2 + m^2}$$

$$\times \left\{ \frac{(\mathbf{p}_\perp^2 + m^2)\,\mathrm{Tr}[(\not{p}_\perp - \not{r}_1 + \not{q}_{1\perp} + m)(-\not{p}_\perp - \not{r}_1 - \not{q}_{1\perp} + m)]}{x(\mathbf{p}_\perp - \mathbf{r}_1 + \mathbf{q}_{1\perp})^2 + (1-x)\mathbf{p}_\perp^2 + m^2} \right.$$

$$\left. - \frac{\mathrm{Tr}[(-\not{p}_\perp - \not{r}_1 - \not{q}_{1\perp} + m)(-\not{q}_{2\perp} + \not{q}_{1\perp} + \not{p}_\perp + m)(-\not{r}_1 + \not{q}_{2\perp} - \not{p}_\perp + m)(\not{p}_\perp + m)]}{(1-x)(\mathbf{p}_\perp + \mathbf{r}_1 - \mathbf{q}_{2\perp})^2 + x(\mathbf{p}_\perp + \mathbf{q}_{1\perp} - \mathbf{q}_{2\perp})^2 + m^2} \right\}$$

$$(11.37)$$

and

$$\mathcal{I}^f = g^2 / 2m. \quad (11.38)$$

Thus the eighth-order amplitude for fermion-fermion scattering is asymptotically of the order of $s \ln s$, as is explicitly given in relation (11.36).

11.6   Tower Diagrams

The existence of uncanceled logarithms in the eighth-order amplitude for fermion-fermion scattering is the first evidence that the total cross section does not approach a constant in the high-energy limit. It does not mean that the total cross section rises as $\ln s$ with the energy. This is because there are $s(\ln s)^n$, $n = 1, 2, 3, \ldots$, terms in the elastic scattering amplitude.

For fermion-fermion scattering the lowest-order amplitude that has uncanceled $s \ln^2 s$ terms is of the twelfth order, with the diagrams contributing to these terms illustrated in figure 11.16. These diagrams are those with two fermion loops joined vertically by two vector mesons. More generally, the lowest-order diagrams that contribute uncanceled $s(\ln s)^n$ terms are those with n fermion loops, each of which is joined vertically to

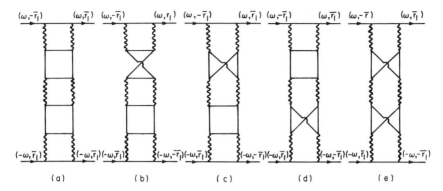

Figure 11.16
The lowest- (twelfth-) order diagrams that yield $s(\ln s)^2$ terms. Note that there are two fermion loops in each diagram.

the adjacent ones by two vector mesons, and they are of $4(n + 1)$th order. These diagrams are called tower diagrams [2, 5].

The calculation of the tower diagrams is straightforward; hence we outline just a few key points. Consider the twelfth-order tower diagrams in figure 11.16. As in the eighth-order tower diagrams, the longitudinal momenta of the fermions and antifermions in the same fermion loop are comparable. Also, the plus (resp. minus) momenta of a fermion loop are much larger than those of another fermion loop below (resp. above) it. Let the minus momenta of the upper and the lower fermion loops in figure 11.16 be Q_1 and Q_2, respectively. Then the amplitude corresponding to the diagrams in figure 11.16 has the factor

$$\int_{1/\omega}^{\omega} \frac{dQ_2}{Q_2} \int_{1/\omega}^{Q_2} \frac{dQ_1}{Q_1} \simeq \frac{1}{2!} \ln^2 s. \tag{11.39}$$

Recall that, although Q_1 in relation (11.39) can be as large as Q_2, the dominant contribution to the integral comes from $Q_1 \ll Q_2$. Relation (11.39) shows how the $s \ln^2 s$ terms come about.

The coefficient of $s \ln^2 s$ can be calculated in the following way. Let the contribution of figures 11.12a and 11.12b to $K(\mathbf{r}_1, \mathbf{q}_{1\perp}, \mathbf{q}_{2\perp})$ be K_1 and that of figure 11.12c to $K(\mathbf{r}_1, \mathbf{q}_{1\perp}, \mathbf{q}_{2\perp})$ be K_2, with

$$K = K_1 + K_2.$$

Then the diagrams in figures 11.16a and b contribute a term K_1^2, that in figure 11.16c a term $K_2 K_1$, that in figure 11.16d a term $K_1 K_2$, and that in figure 11.16e a term K_2^2. Some explanation of the combinatorial factors involved is in order. Figure 11.16a has only one flow diagram; so does figure 11.16b. These two flow diagrams differ only by the propagator of the top fermion line, and the sum of the two corresponding amplitudes is equal to the amplitude of figure 11.16a with the replacement of

$$\frac{1}{Q + i\varepsilon} \rightarrow \frac{1}{Q + i\varepsilon} + \frac{1}{-Q} = -i\pi\delta(Q),$$

where Q is the minus momentum of the top fermion line. The diagram in figure 11.16c has two flow diagrams, which differ only in the direction of the arrow on the top fermion line. Again, the sum of the two corresponding amplitudes gives the factor $-i\pi\delta(Q)$. Figure 11.16d can be discussed in the same way as figure 11.16c. Finally, figure 11.16e has four flow diagrams that can be grouped into two pairs. Each pair of flow diagrams gives a factor $-i\pi\delta(Q)$; thus the two pairs of diagrams give a factor $-2i\pi\delta(Q)$. According to the Feynman rules, however, there is an extra combinatorial factor $\frac{1}{2}$ for the amplitude of the diagram in figure 11.16e, as this diagram is invariant on interchanging the roles of the two vector mesons connecting the two loops (see figure 11.17).

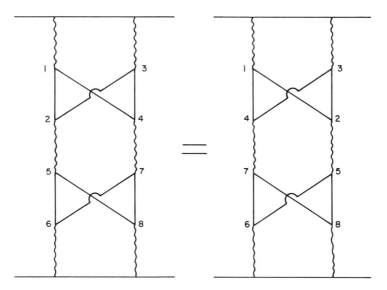

Figure 11.17
Two ways to draw figure 11.16e.

The calculations for the $4(n + 1)$th-order tower diagrams can be carried out in a similar way. Using identities (6.16) and (6.17), we get

$$\mathcal{M}_n \simeq \frac{is(\ln s)^n}{n!}(J, \mathcal{K}_t^n J),\tag{11.40}$$

as given in relation (6.21). For $s \to \infty$, this leading term is larger for larger n. It is therefore natural to sum relation (11.40) over n:

$$\sum_{n=0}^{\infty} i(n!)^{-1}s(\ln s)^n(J, \mathcal{K}_t^n J) = is(J, e^{\mathcal{K}_t \ln s}J)$$

$$= is(J, s^{\mathcal{K}_t}J).\tag{11.41}$$

We recall from equation (11.37) that the operator K depends on the momentum transfer $\Delta = 2\mathbf{r}_1$. Note the formal similarity between equation (11.41) and the amplitude resulting from a Regge pole [6]. Indeed equation (11.41) is more general: If $\mathcal{K}_t$ were not an operator but a function that depends on Δ, then we would get a Regge pole. If $\mathcal{K}_t$ were a Fredholm operator (which necessarily has a discrete spectrum), then we would get a series of Regge poles. For this reason it is important to find the spectrum of $\mathcal{K}_t$. This will also enable us to evaluate equation (11.41).

As a first step in this direction, we cast equation (11.37) in a different form. By carrying out the integration over x and replacing $\mathbf{p}_\perp$ by $\mathbf{p}_\perp - \mathbf{q}_\perp$, we get

$$K(\mathbf{r}_1, \mathbf{q}_\perp, \mathbf{q}_\perp') = g^4(2\pi)^{-4}\int d^2p_\perp \int_0^1 dx[x(\mathbf{p}_\perp + \mathbf{r}_1)^2 + (1-x)(\mathbf{p}_\perp - \mathbf{q}_\perp)^2 + m^2]^{-1}$$

$$\times \left\{ -\frac{[(\mathbf{p}_\perp - \mathbf{q}_\perp)^2 + m^2]\operatorname{Tr}(\not{p}_\perp + \not{r}_1 - m)(\not{p}_\perp - \not{r}_1 + m)}{x(\mathbf{p}_\perp - \mathbf{r}_1)^2 + (1-x)(\mathbf{p}_\perp - \mathbf{q}_\perp)^2 + m^2} \right.$$

$$\left. -\frac{\operatorname{Tr}(\not{p}_\perp + \not{r}_1 - m)(\not{p}_\perp - \not{q}_\perp' + m)(\not{p}_\perp - \not{q}_\perp - \not{q}_\perp' + \not{r}_1 - m)(\not{p}_\perp - \not{q}_\perp + m)}{x(\mathbf{p}_\perp - \mathbf{q}_\perp)^2 + (1-x)(\mathbf{p}_\perp - \mathbf{q}_\perp - \mathbf{q}_\perp' + \mathbf{r}_1)^2 + m^2} \right\}$$

$$= 4g^4(2\pi)^{-4}\int d^2p_\perp \left\{ \frac{\mathbf{p}_\perp^2 - \mathbf{r}_1^2 + m^2}{(\mathbf{p}_\perp + \mathbf{r}_1)^2 - (\mathbf{p}_\perp - \mathbf{r}_1)^2}\ln\frac{(\mathbf{p}_\perp + \mathbf{r}_1)^2 + m^2}{(\mathbf{p}_\perp - \mathbf{r}_1)^2 + m^2} \right.$$

$$-\frac{1}{4}\frac{\operatorname{Tr}(\not{p}_\perp + \not{r}_1 - m)(\not{p}_\perp - \not{q}_\perp' + m)(\not{p}_\perp - \not{q}_\perp - \not{q}_\perp' + \not{r}_1 - m)(\not{p}_\perp - \not{q}_\perp + m)}{[(\mathbf{p}_\perp + \mathbf{r}_1)^2 + m^2][(\mathbf{p}_\perp - \mathbf{q}_\perp - \mathbf{q}_\perp' + \mathbf{r}_1)^2 + m^2] - [(\mathbf{p}_\perp - \mathbf{q}_\perp)^2 + m^2][(\mathbf{p}_\perp - \mathbf{q}_\perp')^2 + m^2]}$$

$$\left. \times \ln\frac{[(\mathbf{p}_\perp + \mathbf{r}_1)^2 + m^2][(\mathbf{p}_\perp - \mathbf{q}_\perp - \mathbf{q}_\perp' + \mathbf{r}_1)^2 + m^2]}{[(\mathbf{p}_\perp - \mathbf{q}_\perp)^2 + m^2][(\mathbf{p}_\perp - \mathbf{q}_\perp')^2 + m^2]} \right\}.\tag{11.42}$$

We note that the K of equation (11.42) satisfies the properties

$$K(\mathbf{r}_1, \mathbf{q}_\perp, \mathbf{q}'_\perp) = K(\mathbf{r}_1, \mathbf{q}'_\perp, \mathbf{q}_\perp), \tag{11.43}$$

$$K(\mathbf{r}_1, \pm\mathbf{r}_1, \mathbf{q}'_\perp) = K(\mathbf{r}_1, \mathbf{q}_\perp, \pm\mathbf{r}_1) = 0. \tag{11.44}$$

By equations (11.42) and (11.43), $\mathscr{K}_t$ is real and symmetric. Thus the spectrum $\mathscr{S}(\Delta)$ of $\mathscr{K}_t$ is a real set. In terms of this notation, we show in appendix B that the lowest upper bound of $\mathscr{S}(\Delta)$ is

$$\tfrac{11}{32}\alpha^2\pi, \tag{11.45}$$

independent of Δ. Here $\alpha = g^2/4\pi$. Furthermore this point (11.45) is not a discrete eigenvalue. In Regge language this means that the leading singularity in the complex angular momentum plane is located at $1 + (11/32)\alpha^2\pi$, which is larger than 1 [7].

As a consequence, the high-energy behavior given by equation (11.41) for fermion-fermion elastic scattering is not that of a Regge pole but that of a Regge cut. The result has been given by expression (6.23), and we outline the derivation of this result in appendix B.

References

[1] H. Cheng and T. T. Wu, *Physical Review Letters* 22 (1969):666, *Physical Review* 182 (1969):1852, 1868, 1873, 1899.

[2] See reference [1] of chapter 6.

[3] H. Cheng and T. T. Wu, *Physical Review* 184 (1969):1868; Y. P. Yao, *Physical Review D* 1 (1970):2316, 2971, 3 (1971):1364; S. J. Chang, *Physical Review D* 1 (1970):2977; B. J. Kirby and H. M. Fried, *Physical Review D* 8 (1973):2668.

[4] M. Greco, *Nuovo Cimento A* 4 (1971):689; E. A. Kuraev, L. N. Lipatov, and N. P. Merenkov, *Soviet Journal of Nuclear Physics* 18 (1974):554.

[5] H. M. Fried and C. I. Tan, *Physical Review D* 9 (1974):314; L. Matsson, *Physical Review D* 10 (1974):2010, 2027; M. Davidson, *Physical Review D* 10 (1974):2464; A. M. Carroll, *Annals of Physics* 107 (1977):283; H. M. Fried, *Physical Review D* 24 (1981):1997; G. Alberi and G. Goggi, *Physics Reports* 74 (1981):1.

[6] R. A. Brandt and M. Feinroth, *Physical Review* 187 (1969):1888; S. J. Chang and T. M. Yan, *Physical Review Letters* 25 (1970):1586, *Physical Review D* 4 (1971):537, 10 (1974):1531; B. Hasslacher, D. K. Sinclair, G. M. Cicuta, and R. L. Sugar, *Physical Review Letters* 25 (1970):1591; G. M. Cicuta and R. L. Sugar, *Physical Review D* 3 (1971):970; B. Hasslacher and D. K. Sinclair, *Physical Review D* 3 (1971):1770; S. J. Chang, T. M. Yan, and Y. P. Yao, *Physical Review D* 4 (1971):3012, 5 (1972):271; J. B. Kogut, *Physical Review D* 4 (1971):3101; P. Nicoletopoulos and M. A. L. Prevost, *Nuovo Cimento A* 4 (1971):25, 5 (1971):357, 13 (1973):43; A. R. Swift, *Physical Review D* 5 (1972):1400; J. Bartels, *Annals of Physics* 94 (1975):1, 19; C. G. Callan and M. L. Goldberger, *Physical Review D* 11 (1975):1553; W. Czyż, and P. K. Kabir, *Physical Review D* 11 (1975):2219; G. Calucci, R. Jengo, and V. Roberto, *Nuclear Physics B* 133 (1978):461.

[7] Y. S. Kim and N. Kwak, *Lettere al Nuovo Cimento* 4 (1970):604; J. L. Cardy, *Nuclear Physics B* 75 (1974):413, 93 (1975):525; N. S. Craigie and G. Preparata, *Physics Letters B* 52 (1974):84; R. O. Mastalir, *Journal of Mathematical Physics* 16 (1975):743; B. M. Barbashov and V. V. Nesterenko, *Soviet Journal of Nuclear Physics* 20 (1975):114; V. D. Mur and V. S. Popov, *Soviet Journal of Nuclear Physics* 21 (1975):447; I. J. Muzinich and H. S. Tsao, *Physical Review D* 11 (1975):2203; B. A. Arbuzov, V. Yu. D'yakonov, and V. E. Rochev, *Soviet Journal of Nuclear Physics* 23 (1976):475; P. E. Volkovitskii, A. M. Lapidus, V. I. Lisin, and K. A. Ter-Martirosyan, *Soviet Journal of Nuclear Physics* 24 (1976):648; B. Z. Kopeliovich and L. I. Lapidus, *Soviet Physics JETP* 44 (1976):31; L. Caneschi, *Nuclear Physics B* 108 (1976):417; P. E. Volkovitskii and A. M. Lapidus, *Soviet Journal of Nuclear Physics* 31 (1980):250, 380.

12 Quark-Quark Scattering in Non-Abelian Gauge Field Theories

12.1 Introduction

In the preceding chapter we discussed in great detail the high-energy amplitude of fermion-fermion scattering in the Abelian gauge field theory. In this chapter we repeat the same diagrammatic study for the non-Abelian gauge field theory [1–4].

To be specific, we take the symmetry group to be that of $SU(3)$, although all our results can be easily generalized to other groups. (For example, if the symmetry group is $SU(2)$, we simply replace f_{abc} in the equations in this chapter by ε_{abc}.) We refer to the fermion and the vector meson in this theory as the quark and the gluon, respectively. For most of this chapter we restrict ourselves to quark-quark scattering. Generalizations to gluon-gluon scattering and quark-gluon scattering are almost trivial and are discussed in section 12.5.

The quark-quark scattering amplitude is more complicated than the fermion-fermion scattering amplitude discussed in chapter 11 in one aspect: The quark carries color; hence more amplitudes are required to describe the various color-changing amplitudes as well as the non-color-changing amplitude. It is gratifying, therefore, that a pattern of regularity exists, allowing us to identify each term of the quark-quark scattering amplitude as the counterpart of a term in the fermion-fermion scattering amplitude. Indeed, the main qualitative difference between high-energy QCD and high-energy QED is that the gluon is Reggeized [5], whereas the photon is not.

Because the gluon is massless, the quark-quark scattering amplitude, as with the electron-electron scattering amplitude in QED, has infrared divergences. These divergences are implicitly contained in the transverse-momentum integrals that appear in the (s-independent) coefficients of the $s \ln^n s$ terms. To make the results rigorous, we could give the gluon a mass λ by means of the Higgs mechanism [6]; however, no new features of high-energy scattering are obtained in such a treatment [7]. For the sake of simplicity, we therefore keep the gluon massless and ignore the infrared divergences. Indeed, for the scattering of color singlets, infrared divergences do not occur [8].

12.2 Second and Fourth Orders

There are two second-order diagrams for quark-quark scattering. One of them is illustrated in figure 12.1, and the other can be obtained from

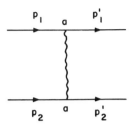

Figure 12.1
A second-order diagram for quark-quark scattering.

figure 12.1 by interchanging the two incident particles. Just as in the Abelian gauge field theory, the latter diagram gives an amplitude that is too small by a factor of u/t and is neglected. By applying the Feynman rules in table 12.1, we find that the scattering amplitude corresponding to figure 12.1 asymptotically approaches

$$-\frac{s}{2m^2}\frac{g^2}{\Delta^2}T_a^{(1)}T_a^{(2)}. \tag{12.1}$$

In expression (12.1), $T_a = \lambda_a/2$ is defined by equation (2.52), with $a = 1$, $2, \ldots, 8$. The superscript for T_a refers to the quark it operates on. For example, $T_a^{(1)}$ is understood to be inserted between the color wave functions of the first quark. Similarly for $T_a^{(2)}$. The repeated color index a in expression (12.1) implies summation. We see that expression (12.1) differs from the corresponding fermion-fermion amplitude in the Abelian gauge field theory only by the group factor $T_a^{(1)}T_a^{(2)}$. For convenience, we refer to the amplitude with the group factor omitted as the space-time factor. The space-time factor for a diagram without three-gluon-coupling vertices in a non-Abelian gauge field theory is the same as the amplitude for the same diagram in the Abelian gauge field theory; however, the existence of the group factor complicates the calculation. Specifically, when we sum amplitudes in the Abelian gauge field theory, we need to add only the space-time factors. But when we sum amplitudes in a non-Abelian gauge field theory, this is no longer the case.

The existence of the group factor leads to new considerations, which we illustrate by studying the box and the crossed box diagrams of figure 12.2. The asymptotic forms of the space-time factors for these diagrams were given in chapter 10. Referring to expression (10.14), we find that the amplitudes corresponding to figures 12.2a and 12.2b are, respectively,

Table 12.1
Feynman rules for QCD[a]

Diagram	Rule
  Internal gluon line	$-\dfrac{ig^{\mu\nu}\delta_{ab}}{k^2 + i\varepsilon}$
  Internal quark line	$\dfrac{i}{\not{p} - m + i\varepsilon}$
  Quark-quark-gluon vertex	$-ig\gamma_\mu \lambda_a/2$
  Three gluon vertex	$gf_{abc}[g_{\rho\mu}(q-k)_\nu + g_{\mu\nu}(k-p)_\rho + g_{\mu\rho}(p-q)_\mu]$
  Four gluon vertex	$-ig^2\{f_{abe}f_{cde}(g_{\mu\rho}g_{\nu\sigma} - g_{\mu\sigma}g_{\nu\rho})$ $+ f_{ace}f_{bde}(g_{\mu\nu}g_{\rho\sigma} - g_{\mu\rho}g_{\nu\sigma})$ $+ f_{ade}f_{bce}(g_{\mu\nu}g_{\rho\sigma} - g_{\mu\rho}g_{\mu\sigma})\}$

a. The other rules are the same as those in QED given in figure 3.1, with the exception that an external line here is associated with an additional wave function in the abstract space of internal symmetry. For the case of $SU(2)$, replace f_{abc} by ε_{abc} and λ_a by σ_a.

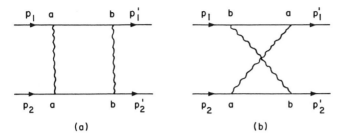

Figure 12.2
The box (a) and the crossed box (b) diagrams in quark-quark scattering.

$$\mathcal{M}\, T_b^{(1)} T_a^{(1)} T_b^{(2)} T_a^{(2)},$$

$$\mathcal{M}'\, T_a^{(1)} T_b^{(1)} T_b^{(2)} T_a^{(2)},\tag{12.2a}$$

where

$$\mathcal{M} \simeq -\frac{1}{2} g^4 \frac{s}{m^2}\frac{\ln(se^{-i\pi})}{2\pi} I(\mathbf{\Delta}),$$

$$\mathcal{M}' \simeq \frac{1}{2} g^4 \frac{s}{m^2}\frac{\ln s}{2\pi} I(\mathbf{\Delta}),\tag{12.2b}$$

and

$$I(\mathbf{\Delta}) \equiv \int \frac{d^2 q_\perp}{(2\pi)^2}\frac{1}{(\mathbf{q}_\perp^2)[(\mathbf{\Delta} - \mathbf{q}_\perp)^2]}.\tag{12.3}$$

In relations (12.2b) both the leading real parts and the leading imaginary parts of $\mathcal{M}$ and $\mathcal{M}'$ are included. As we have mentioned, $I(\mathbf{\Delta})$ is an infrared divergent integral. If we invoke the Higgs mechanism, then $(\mathbf{q}_\perp^2)[(\mathbf{\Delta} - \mathbf{q}_\perp)^2]$ in identity (12.3) would be replaced by $(\mathbf{q}_\perp^2 + \lambda^2)[(\mathbf{\Delta} - \mathbf{q}_\perp)^2 + \lambda^2]$, and the integral becomes convergent.

Because the group factors associated with $\mathcal{M}$ and $\mathcal{M}'$ are different, it is convenient to put

$$T_b^{(1)} T_a^{(1)} = \tfrac{1}{2}[T_b^{(1)}, T_a^{(1)}]_+ + \tfrac{1}{2}[T_b^{(1)}, T_a^{(1)}],\tag{12.4}$$

$$T_a^{(1)} T_b^{(1)} = \tfrac{1}{2}[T_b^{(1)}, T_a^{(1)}]_+ - \tfrac{1}{2}[T_b^{(1)}, T_a^{(1)}],\tag{12.5}$$

where

$$[T_b, T_a]_+ \equiv T_b T_a + T_a T_b.$$

The commutator can be simplified by using the commutation rule (2.53):

$$[T_a, T_b] = if_{abc} T_c, \tag{12.6}$$

where f_{abc} is the structure constant of the symmetry group. By substituting equations (12.4) and (12.5) into expression (12.2a), we find that the sum of the amplitudes corresponding to the diagrams in figure 12.2 is

$$\mathscr{M}_+ [T_a^{(1)}, T_b^{(1)}]_+ [T_a^{(2)}, T_b^{(2)}]_+ + \mathscr{M}_- [T_a^{(1)}, T_b^{(1)}][T_a^{(2)}, T_b^{(2)}], \tag{12.7}$$

where

$$\mathscr{M}_+ = \frac{1}{4}(\mathscr{M} + \mathscr{M}') \simeq \frac{g^4}{16} \frac{is}{m^2} I(\Delta), \tag{12.8}$$

$$\mathscr{M}_- = \frac{1}{4}(\mathscr{M} - \mathscr{M}') \simeq -\frac{g^4 s}{8\pi m^2} \ln(se^{-i\pi/2}) I(\Delta). \tag{12.9}$$

By applying equation (12.6), we can reduce the sum of the two amplitudes corresponding to the diagrams in figure 12.2 to

$$\mathscr{M}_+ [T_a^{(1)}, T_b^{(1)}]_+ [T_a^{(2)}, T_b^{(2)}]_+ - 3\mathscr{M}_- T_c^{(1)} T_c^{(2)}, \tag{12.10}$$

where we have made use of the formula

$$f_{abc} f_{abc'} = 3\delta_{cc'}. \tag{12.11}$$

The point to note in equation (12.8) is that $\mathscr{M}_+$ is obtained by adding the space-time factors of $\mathscr{M}$ and $\mathscr{M}'$, which is the way we add amplitudes in the Abelian gauge field theory. Therefore, just like its counterpart in the Abelian gauge field theory, $\mathscr{M}_+$ is imaginary and is of magnitude $g^4 s$. The amplitude $\mathscr{M}_-$ is obtained by taking the difference of the space-time factors of $\mathscr{M}$ and $\mathscr{M}'$. This amplitude has no counterpart in the Abelian gauge field theory. It is predominantly real and is as large as $sg^4 \ln s$, that is, larger than $\mathscr{M}_+$ by a factor $\ln s$. Mathematically, $\mathscr{M}_-$ is present because the operators T_a do not commute with one another. From the physical viewpoint, the quantity $\mathscr{M}_-$ is a color-changing amplitude, as evidenced by the fact that the group factor associated with $\mathscr{M}_-$ is the same as that associated with the amplitude (12.1), which describes the exchange of a color gluon.

The other fourth-order diagrams for quark-quark scattering are either too small by a factor s/t or just dress the quark form factor and the gluon propagator.

Before we close this section, let us discuss a diagrammatic method to treat the group factor. We define the group factor of a diagram as the matrix in the color space obtained by associating with each vertex in the diagram the factor shown in figure 12.3. Instead of writing it explicitly, however, we represent the group factor associated with a Feynman diagram by precisely that diagram. As an example, the group factor $T_a^{(1)}T_a^{(2)}$ is represented by the diagram in figure 12.1, whereas the group factors $T_b^{(1)}T_a^{(1)}T_b^{(2)}T_a^{(2)}$ and $T_a^{(1)}T_b^{(1)}T_b^{(2)}T_a^{(2)}$ are represented by the diagrams in figures 12.2a and 12.2b, respectively. In this notation, the commutation rule (12.6) is simply expressed by the diagrammatic equation in figure 12.4a. Similarly, the Jacobi identity for the structure constant

$$f_{abe}f_{cde} + f_{ace}f_{dbe} + f_{ade}f_{bce} = 0 \tag{12.12}$$

is diagrammatically represented by figure 12.4b, which is just figure 12.4a with the quark line replaced by a gluon line. Note that the structure constant f_{abc} changes sign as we make an odd permutation of its indexes. Hence the term on the right-hand side of figure 12.4c differs in sign from that of figure 12.4a. Another useful relation between the group factors can be obtained by multiplying equation (12.6) by $f_{abc'}$ and making use of equation (12.11). We get

$$2f_{abc}T_aT_b = 3iT_c,$$

or

$$if_{cba}T_aT_b = \tfrac{3}{2}T_c. \tag{12.13}$$

Equation (12.13) is diagrammatically represented by figures 12.5a and 12.5b. (Note the difference of the signs in the right-hand side of these equations.) We can replace the quark line in figure 12.5a by a gluon line, obtaining figure 12.5c. The latter diagrammatic equality can be proved by multiplying equation (12.12) by $f_{abc'}$. There is no change of sign for the right-hand side of figure 12.5c if we turn the diagrams upside down. This is because both diagrams in figure 12.5c contain odd numbers (3 and 1, respectively) of factors of the structure constant, and the equation does not change if all structure constants change sign.

From the diagrammatic equations figures 12.4 and 12.5, we can obtain the diagrammatic equation represented by figure 12.6. Thus the sum of the two amplitudes corresponding to the diagrams in figure 12.2 and given by equation (12.2) is simply

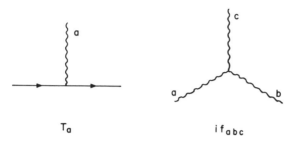

Figure 12.3
The group factors for the quark-quark-gluon vertex and the three-gluon vertex.

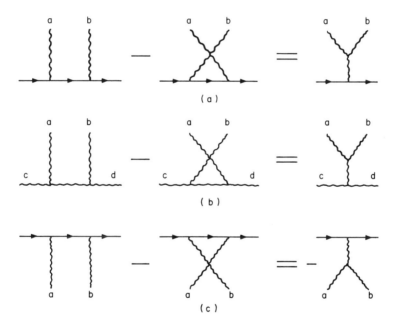

Figure 12.4
Diagrammatic equations for the group factors.

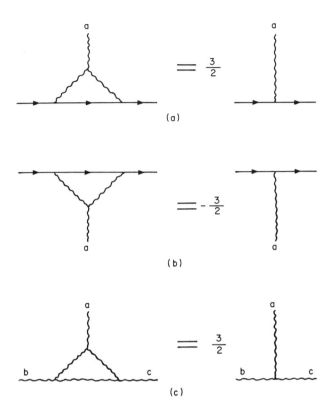

Figure 12.5
Further diagrammatic equations for the group factors.

$$(\mathcal{M} + \mathcal{M}')G_2 + \tfrac{3}{2}\mathcal{M}'G_1, \tag{12.14}$$

where G_1 and G_2 are group factors defined in figure 12.7 and where, from relation (12.2b),

$$\mathcal{M} + \mathcal{M}' \simeq \frac{ig^4}{4} \frac{s}{m^2} I(\Delta),$$

and

$$\frac{3}{2}\mathcal{M}' \simeq \frac{3}{4}g^4 \frac{s}{m^2} \frac{\ln s}{2\pi} I(\Delta). \tag{12.15}$$

By adding the second-order amplitude (12.1) to the fourth-order amplitude (12.14), we find that the scattering amplitude up to the fourth order

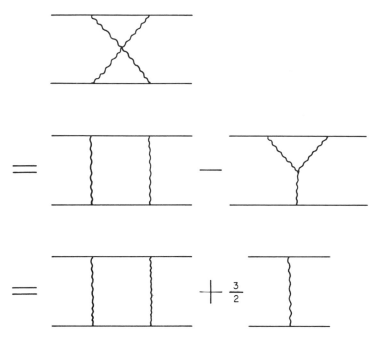

Figure 12.6
Another diagrammatic equation for the group factors.

asymptotically approaches

$$-\frac{s}{2m^2}\frac{g^2}{\Delta^2}[1-\bar{\alpha}(\Delta)\ln s]G_1+\frac{ig^4}{4}\frac{s}{m^2}I(\Delta)G_2,\qquad(12.16)$$

where

$$\bar{\alpha}(\Delta)=\frac{3g^2}{4\pi}\Delta^2 I(\Delta).\qquad(12.17)$$

We make the following comments.

1. As we have mentioned, G_1 is a color-changing matrix. More precisely, the quantum numbers of G_1 in the t-channel are those of a gluon, which is a color octet. Similarly, the quantum numbers of G_2 in the t-channel are those of two gluons, a state of 8×8 in the color space. Such a state is a superposition of several irreducible representations of $SU(3)$, one of which is the color singlet. We see from expression (12.16) that up to the fourth order, the color-changing amplitude is larger than the vacuum amplitude

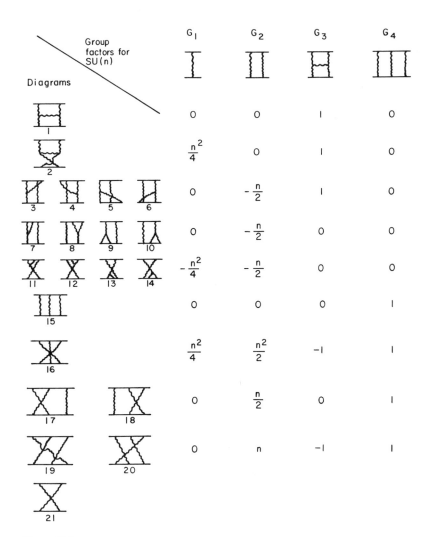

Figure 12.7
Relation between G_i, $i = 1, 2, 3, 4$, and the group factor associated with a Feynman diagram.
(Each G_i is defined by the diagram below it.) For example, the group factor for diagram 2 in
this figure is equal to $(n^2/4)G_1 + G_3$.

(amplitude with no exchange of quantum numbers) by a factor of $\ln s$. Actually this will no longer be true if all leading terms in the perturbation series are summed. As we will see, the terms in the coefficients of G_1 add up to an exponential:

$$-\frac{s}{2m^2}\frac{g^2}{\Delta^2}\left[1 - \bar{\alpha}(\Delta)\ln s + \frac{\bar{\alpha}^2(\Delta)\ln^2 s}{2!} - \cdots\right] = -\frac{g^2}{2m^2}\frac{s^{\alpha(\Delta)}}{\Delta^2}, \qquad (12.18)$$

where

$$\alpha(\Delta) \equiv 1 - \frac{3g^2}{4\pi}\Delta^2 I(\Delta). \qquad (12.19)$$

The amplitude (12.18) is a Regge pole term. Because $\alpha < 1$, the higher-order terms depress the amplitude.

2. If we introduce Higgs bosons to give a mass λ to the gluons, then relations (12.18) and (12.19) are modified:

$$-\frac{g^2}{2m^2}\frac{s^{\alpha(\Delta)}}{\Delta^2 + \lambda^2}, \qquad (12.20)$$

where

$$\alpha(\Delta) \equiv 1 - \frac{3g^2}{4\pi}(\Delta^2 + \lambda^2)\int\frac{d^2q_\perp}{(2\pi)^2}\frac{1}{(\mathbf{q}_\perp^2 + \lambda^2)[(\Delta - \mathbf{q}_\perp)^2 + \lambda^2]}. \qquad (12.21)$$

There are no infrared divergences in expression (12.20) and identity (12.21).

3. If the symmetry group is $SU(n)$, expression (12.16) still holds, with the factor 3 in equation (12.17) replaced by n.

4. In expression (12.16) the terms neglected in the coefficient of G_1 are of the order of $g^4 s$. Such terms modify the gluon propagator and the coupling constant. The terms neglected in the coefficient of G_2 are of the order of $g^4 s^0$ (logarithms ignored).

In summary, the only qualitative differences in expression (12.16) with the second- through fourth-order amplitudes in Abelian gauge theory are the existence of the group factors and the Reggeization of the gluon.

12.3   Sixth Order

For quark-quark scattering there are twenty-one sixth-order diagrams that contribute to $s\ln^2 s$ and $s\ln s$ terms. These diagrams are illustrated in figure 12.7. In this figure the group factors for these diagrams are decomposed as linear combinations of G_i, $i = 1, 2, 3, 4$. This is easily derived from the diagrammatic equations of figures 12.4 and 12.5.

Before we go into the details of the calculation, we mention that the amplitudes $\mathscr{M}_1$, $\mathscr{M}_2$, and $\mathscr{M}_{21}$, corresponding to the diagrams designated as 1, 2, and 21 in figure 12.7, are as large as s^2. However, these s^2 terms cancel as they are summed. We observe that diagram 21 is obtained from diagram 1 or diagram 2 by fusing the middle rung. As a general rule, s^2 terms cancel as we add the fused diagram to the original one. It is therefore simpler if we add the amplitudes of the fused diagrams to those of the original diagram *before* we make asymptotic calculations. It is to be noted that there are three original diagrams that lead to the same fused diagram. The Feynman rule for a four-particle vertex also contains three terms, and we should associate each of these terms with the corresponding original diagram. (See the Feynman rule for the four-gluon vertex in table 12.1.) The existence and cancellation of the s^2 term parallel the situation with the tower diagrams in the Abelian gauge field theory discussed in section 11.6.

Diagrams 1 and 2 (figure 12.7)

The scattering amplitude corresponding to diagram 1 in figure 12.7, with the numerator set to unity, is given by expression (10.24b). Let us calculate the numerator for the space-time factor of diagram 1 in figure 12.7. Just as in the Abelian case discussed in chapter 3, the contribution of the quark line at the top to the numerator factor is approximately

$2p_{1\mu}p_{1\nu}/m,$

where μ and ν are, respectively, the polarizations of line 4 and line 6 in figure 12.8. Similarly for the contribution of the quark line at the bottom. Thus the numerator for the diagram in figure 12.8 is approximately

$$N \simeq \frac{4\omega^4}{m^2} \{2[(q_1 - 2q_2)_+ + (q_2 + q_1)_+][(q_2 - 2q_1)_- + (q_1 + q_2)_-]$$

$$+ 2[(q_1 + q_2)_- + (q_2 - 2q_1)_-][(q_1 - 2q_2)_+ + (q_2 + q_1)_+]$$

$$- 4(\mathbf{q}_{1\perp} + \mathbf{q}_{2\perp})\cdot(\mathbf{q}_{1\perp} + \mathbf{q}_{2\perp} - 2\mathbf{\Delta})\}.$$

We recall that the dominant contribution comes from the region where

$q_{1+} \simeq q_{2-} \simeq 0.$

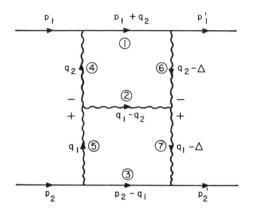

Figure 12.8
The sixth-order ladder diagram for quark-quark scattering. The plus and minus signs represent the polarizations of the corresponding lines.

Thus

$$N \simeq \frac{16\omega^4}{m^2} [q_{2+}q_{1-} - (\mathbf{q}_{1\perp} + \mathbf{q}_{2\perp}) \cdot (\mathbf{q}_{1\perp} + \mathbf{q}_{2\perp} - 2\boldsymbol{\Delta})]. \tag{12.22}$$

As we have mentioned, we have to add part of the contribution of diagram 21 (figure 12.7) to this N. According to the Feynman rules, it is

$$\frac{16\omega^4}{m^2}(q_1 - q_2)^2 \simeq \frac{16\omega^4}{m^2}[-q_{2+}q_{1-} - (\mathbf{q}_{1\perp} - \mathbf{q}_{2\perp})^2]. \tag{12.23}$$

Adding the quantity above to N in relation (12.22) and denoting this sum by $\bar{N}$, we get

$$\bar{N} \simeq -\frac{16\omega^4}{m^2}[(\mathbf{q}_{1\perp} + \mathbf{q}_{2\perp}) \cdot (\mathbf{q}_{1\perp} + \mathbf{q}_{2\perp} - 2\boldsymbol{\Delta}) + (\mathbf{q}_{1\perp} - \mathbf{q}_{2\perp})^2]$$

$$= -\frac{32\omega^4}{m^2}[\mathbf{q}_{1\perp}^2 + \mathbf{q}_{2\perp}^2 - \boldsymbol{\Delta} \cdot (\mathbf{q}_{1\perp} + \mathbf{q}_{2\perp})]. \tag{12.24}$$

By combining relations (10.24a) and (12.24), we find that the amplitude corresponding to diagram 1 (plus the contribution from diagram 21) is

$$-\frac{g^6}{4} \frac{s}{m^2} \left[\frac{\ln(se^{-i\pi})}{2\pi}\right]^2 \int \prod_{i=1}^{2} \frac{d^2 q_{i\perp}}{(2\pi)^2} \frac{\mathbf{q}_{1\perp}^2 + \mathbf{q}_{2\perp}^2 - \boldsymbol{\Delta} \cdot (\mathbf{q}_{1\perp} + \mathbf{q}_{2\perp})}{\prod_{i=1}^{2}[(\mathbf{q}_{i\perp}^2)(\boldsymbol{\Delta} - \mathbf{q}_{i\perp})^2]}. \tag{12.25}$$

(An additional minus sign has to be added to relation (10.24a), as the middle rung of diagram 1 is not a scalar meson but a gluon, the propagator of which has an additional minus sign.) The term $\Delta \cdot \mathbf{q}_{11\perp}$ in the numerator in equation (12.25) can be symmetricized into

$$\Delta \cdot \mathbf{q}_{11\perp} \rightarrow \Delta \cdot [\mathbf{q}_{11\perp} + (\Delta - \mathbf{q}_{11\perp})]/2 = \tfrac{1}{2}\Delta^2,$$

as the denominator in expression (12.25) is invariant under the change of variable

$$\mathbf{q}_{11\perp} \rightarrow \Delta - \mathbf{q}_{11\perp}.$$

Similarly, $\Delta \cdot \mathbf{q}_{21\perp} \rightarrow \tfrac{1}{2}\Delta^2$. Thus the amplitude corresponding to diagram 1 plus the contribution from diagram 21, denoted by $\overline{\mathcal{M}}_1$, is

$$\overline{\mathcal{M}}_1 \simeq -\frac{g^6 s}{m^2}\frac{1}{2}\left[\frac{\ln(se^{-i\pi})}{2\pi}\right]^2 I(\Delta)[K(\Delta) - \tfrac{1}{2}\Delta^2 I(\Delta)], \tag{12.26}$$

where $I(\Delta)$ is defined in identity (12.3) and

$$K(\Delta) \equiv \int \frac{d^2 q_\perp}{(2\pi)^2}\frac{1}{\mathbf{q}_\perp^2}, \tag{12.27}$$

which is ultraviolet divergent, that is, divergent even if a mass is introduced for the gluon. Such divergent quantities are to be canceled by contributions from other diagrams.

The amplitude corresponding to diagram 2 plus the remainder of the contribution from diagram 21 can be obtained from relation (12.26) by setting $s \rightarrow u \simeq -s$. It is

$$\overline{\mathcal{M}}_2 \simeq \frac{g^6 s}{m^2}\frac{1}{2}\left[\frac{\ln s}{2\pi}\right]^2 I(\Delta)[K(\Delta) - \tfrac{1}{2}\Delta^2 I(\Delta)]. \tag{12.28}$$

Diagrams 3–6 (figure 12.7)

The scattering amplitudes from the Feynman diagrams 3, 4, 5, and 6 of figure 12.7 are equal. This is because they are obtained from one another by reflections with respect to either a vertical line or a horizontal line and are hence related either by a time reflection or by a space reflection.

As a consequence, we need to calculate only diagram 3. This Feynman diagram has only one flow diagram, which we illustrate in figure 12.9. According to the discussion following expression (10.34), we can ignore the contribution of a pole on a quark line at the top. Thus we need to take

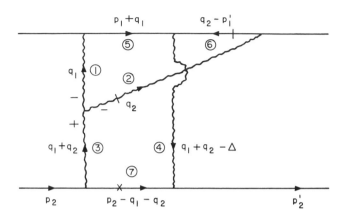

Figure 12.9
The flow diagram corresponding to diagram 3 in figure 12.7. The cross on line 7 means that the pole associated with line 7 is enclosed. The bars on lines 2 and 6 mean that the poles of both these lines are enclosed as the integration is carried out over the plus-momentum of the loop composed of lines 1, 2, 6, and 5.

only the pole of line 2 for the loop composed of lines 1, 2, 6, and 5 and the pole of line 7 for the loop composed of lines 3, 1, 5, 4, and 7. We then have (see equation (10.5))

$$q_{2+} \simeq -q_{1+} \simeq \frac{a_2}{Q_2},$$ (12.29)

where

$$Q_i \equiv q_{i-}, \qquad i = 1, 2.$$

The D function for figure 12.9 is therefore

$$D \simeq \left(-Q_1 \frac{a_2}{Q_2} - a_1\right)(Q_2)(-a_3)(-a_4)(2\omega Q_1 + i\varepsilon)(-2\omega Q_2)(2\omega).$$ (12.30)

The numerator of the Feynman integral corresponding to figure 12.9 is obtained by replacing the vertices on the quark lines at the top and the bottom by

$$\frac{4p_{1\mu}p_{1\nu}p_{1\rho}}{m}\frac{2p_{2\sigma}p_{2\delta}}{m}.$$

Thus we have

$$N \simeq \frac{16\omega^5}{m^2}[2(2Q_1 + Q_2) - 2(Q_1 + 2Q_2)] = \frac{32\omega^5}{m^2}(Q_1 - Q_2). \qquad (12.31)$$

In the region $Q_1 \gg Q_2$ we have from relations (12.30) and (12.31) that

$$\frac{N}{D} \simeq \frac{s}{m^2} \frac{1}{Q_1 Q_2} \frac{1}{a_2 a_3 a_4}.$$

(Because $Q_1 \gg Q_2$, we do not need to keep the $i\varepsilon$ associated with Q_1, in accordance with the discussion following relation (10.58).) The other factors in the scattering amplitude are

$$(-i)(-ig)^6(-i)^4 i^3 \left(-\frac{i}{2}\right)^2 \int \prod_{i=1}^{2} \frac{d^2 q_{i\perp}}{(2\pi)^2} \frac{dQ_i}{2\pi} = -\frac{g^6}{4} \int \prod_{i=1}^{2} \frac{d^2 q_{i\perp}}{(2\pi)^2} \frac{dQ_i}{2\pi},$$

$$(12.32)$$

where $-i$ is the overall factor, $(-ig)^6$ is associated with the order of perturbation, $(-i)^4$ comes from the gluon propagators, i^3 comes from the quark propagators, and $(-i/2)^2$ comes from the two integrations over q_{1+} and q_{2+}. Thus the contribution of the region $Q_1 \gg Q_2$ to the space-time factor of the scattering amplitude is

$$-\frac{s}{4m^2} \frac{g^6}{2} \left(\frac{\ln s}{2\pi}\right)^2 K(\Delta) I(\Delta), \qquad (12.33)$$

where I and K are defined in identities (12.3) and (12.27), respectively. Similarly, in the region $Q_1 \ll Q_2$, we have

$$\frac{N}{D} \simeq -\frac{s}{m^2} \frac{1}{(Q_1 + i\varepsilon)Q_2 a_1 a_3 a_4}.$$

Thus the contribution of the region $Q_1 \ll Q_2$ to the space-time factor of the scattering amplitude is

$$\frac{s}{4m^2} \frac{g^6}{2} \left[\frac{\ln(se^{-i\pi})}{2\pi}\right]^2 K(\Delta) I(\Delta). \qquad (12.34)$$

The asymptotic form of $\mathcal{M}_3$ is equal to the sum of expressions (12.33) and (12.34). Thus we have

$$4\mathcal{M}_3 = \mathcal{M}_3 + \mathcal{M}_4 + \mathcal{M}_5 + \mathcal{M}_6 \simeq \frac{s}{m^2} \frac{g^6}{8\pi^2} [\ln^2(se^{-i\pi}) - \ln^2 s] K(\Delta) I(\Delta).$$

$$(12.35)$$

Diagrams 7–10 (figure 12.7)

The scattering amplitudes from diagrams 7, 8, 9, and 10 of figure 12.7 are equal. Thus we need to calculate only the scattering amplitude from diagram 7. This diagram has only one flow diagram, which is illustrated in figure 12.10. At the poles of line 2 and line 7, we have

$$q_{1+} \simeq -q_{2+} = -\frac{a_2}{Q_2}. \tag{12.36}$$

Thus

$$D \simeq \left(-Q_1\frac{a_2}{Q_2} - a_1\right)(Q_2)(-a_3)(-a_4)(2\omega Q_1 + i\varepsilon)[2\omega(Q_1 + Q_2)](2\omega), \tag{12.37}$$

and

$$N \simeq \frac{32\omega^5}{m^2}(Q_1 - Q_2), \tag{12.38}$$

as in equation (12.31). In the region $Q_1 \gg Q_2$, we have

$$\frac{N}{D} \simeq -\frac{s}{m^2}\frac{1}{Q_1^2 a_2 a_3 a_4}. \tag{12.39}$$

That N/D is proportional to Q_1^{-2} means that the corresponding amplitude

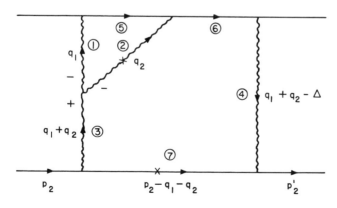

Figure 12.10
The flow diagram corresponding to diagram 7 in figure 12.7. The plus and minus signs represent the polarizations of the corresponding lines.

does not contribute either an $s \ln^2 s$ term or an $is \ln s$ term (it contributes an $s \ln s$ term). Next, we consider the region $Q_1 \ll Q_2$, where

$$\frac{N}{D} \simeq \frac{s}{m^2} \frac{1}{(Q_1 + i\varepsilon)Q_2} \frac{1}{a_1 a_3 a_4}. \tag{12.40}$$

Thus, together with the factor in equation (12.32), we have

$$4\mathcal{M}_7 = \mathcal{M}_7 + \mathcal{M}_8 + \mathcal{M}_9 + \mathcal{M}_{10} \simeq -\frac{s}{m^2} \frac{g^6}{8\pi^2} \ln^2(se^{-i\pi}) K(\Delta) I(\Delta). \tag{12.41}$$

Diagrams 11–14 (figure 12.7)

Diagram 11 in figure 12.7 can be obtained from diagram 8 by interchanging particle 1 with particle 1'. Thus the amplitude corresponding to diagram 11 can be obtained from that corresponding to diagram 8 by making the substitution $s \to u$. Thus we have from equation (12.41) that

$$4\mathcal{M}_{11} = \mathcal{M}_{11} + \mathcal{M}_{12} + \mathcal{M}_{13} + \mathcal{M}_{14} \simeq \frac{s}{m^2} \frac{g^6}{8\pi^2} \ln^2 s \, K(\Delta) I(\Delta). \tag{12.42}$$

Diagrams 15–20 (figure 12.7)

The space-time factors for diagrams 15 through 20 in figure 12.7 with all particles being scalar mesons were calculated on pp. 129–138. For the present case of quark-quark scattering with the exchanged particles being gluons, there is an additional factor of

$$-2(s^3/m^2),$$

where the factor (-1) comes from the fact that a gluon propagator has an additional minus sign $(-ig_{\mu\nu}/q^2)$ and that there are three virtual gluons and where the factor $2s^3/m^2$ comes from the vertices. Thus we have from relations (10.50), (10.52), (10.53), and (10.54) that

$$\mathcal{M}_{15} \simeq \mathcal{M}_{16} \simeq -\frac{g^6 s \ln s}{4\pi^2 m^2} I_2, \tag{12.43}$$

$$\mathcal{M}_{17} = \mathcal{M}_{18} \simeq \frac{g^6 s \ln s}{8\pi^2 m^2} (I_2 + i\pi I_1), \tag{12.44}$$

$$\mathcal{M}_{19} = \mathcal{M}_{20} \simeq \frac{g^6 s \ln s}{8\pi^2 m^2} (I_2 - i\pi I_1), \tag{12.45}$$

where I_1 and I_2 are defined by equations (10.49) and (10.51), respectively.

Summary

According to figure 12.7, the sixth-order scattering amplitude is equal to

$\frac{9}{4}(\bar{\mathcal{M}}_2 - 4\mathcal{M}_{11} + \mathcal{M}_{16})G_1$

$\quad + \frac{3}{2}(-4\mathcal{M}_3 - 4\mathcal{M}_7 - 4\mathcal{M}_{11} + 3\mathcal{M}_{16} + 2\mathcal{M}_{17} + 4\mathcal{M}_{19})G_2$

$\quad + (\bar{\mathcal{M}}_1 + \bar{\mathcal{M}}_2 + 4\mathcal{M}_3 - \mathcal{M}_{16} - 2\mathcal{M}_{19})G_3$

$\quad + (\mathcal{M}_{15} + \mathcal{M}_{16} + 2\mathcal{M}_{17} + 2\mathcal{M}_{19})G_4,$ \hfill (12.46)

the contribution of $\mathcal{M}_{21}$ being included in $\bar{\mathcal{M}}_1$ and $\bar{\mathcal{M}}_2$, as discussed at the beginning of this section.

Consider first the coefficient of G_1. According to relations (12.28), (12.42), and (12.43), we have

$$\frac{9}{4}(\bar{\mathcal{M}}_2 - 4\mathcal{M}_{11} + \mathcal{M}_{16}) \simeq -\frac{9}{8}\frac{g^6 s}{m^2}\frac{1}{2}\left[\frac{\ln s}{2\pi}\right]^2 \Delta^2 I^2(\Delta)$$

$$= -\frac{s}{2m^2}\frac{g^2}{\Delta^2}\frac{[\bar{\alpha}(\Delta)\ln s]^2}{2!},$$ \hfill (12.47)

where $\bar{\alpha}(\Delta)$ is given by equation (12.17). The amplitude $\mathcal{M}_{16}$, being real and of the order of only $s\ln s$ (not $s\ln^2 s$), is neglected in equation (12.47). We observe that the ultraviolet divergent integral K does not appear in equation (12.47). Furthermore, together with expression (12.16), we have found that the coefficient of G_1 is, up to the sixth order,

$$-\frac{s}{2m^2}\frac{g^2}{\Delta^2}\left(1 - \bar{\alpha}(\Delta)\ln s + \frac{\bar{\alpha}^2(\Delta)\ln^2 s}{2}\right),$$

which suggests the validity of equation (12.18).

Next consider the coefficient of G_2. According to equation (12.35) and relations (12.41) through (12.45), we have

$$\frac{3}{2}(-4\mathcal{M}_3 - 4\mathcal{M}_7 - 4\mathcal{M}_{11} + 3\mathcal{M}_{16} + 2\mathcal{M}_{17} + 4\mathcal{M}_{19})$$

$$\simeq -\frac{3is}{m^2}\frac{g^6}{8\pi}\ln s\, I_1(\Delta),$$ \hfill (12.48)

where I_1 is defined in equation (10.49).

Next, according to relations (12.26), (12.28), (12.35), (12.43), and (12.45), the coefficient of G_3 in expression (12.46) is

$$\bar{\mathscr{M}}_1 + \bar{\mathscr{M}}_2 + 4\mathscr{M}_3 - \mathscr{M}_{16} - 2\mathscr{M}_{19} \simeq \frac{is}{m^2} \frac{g^6}{8\pi} \ln s \left[-\Delta^2 I^2(\Delta) + 2I_1(\Delta) \right].$$

$$(12.49)$$

Finally, the coefficient of G_4 is simply the sum of three-vector-meson exchange amplitudes calculated in chapter 3 and given by relation (3.35). Thus the sixth-order amplitude asymptotically approaches

$$\frac{sg^6}{2m^2} \left\{ -\frac{1}{\Delta^2} \frac{[g^{-2}\bar{\alpha}(\Delta)\ln s]^2}{2!} G_1 - \frac{3i}{4\pi} \ln s\, I_1(\Delta) G_2 \right.$$

$$\left. + \frac{i}{4\pi} \ln s [2I_1(\Delta) - \Delta^2 I^2(\Delta)] G_3 + \frac{1}{3!} I_1(\Delta) G_4 \right\}, \quad (12.50)$$

where G_i, $i = 1, 2, 3, 4$, are group factors defined graphically in figure 12.7, and $I(\Delta)$ and $I_1(\Delta)$ are defined by identity (12.3) and equation (10.49), respectively.

12.4   Discussion

Scrutiny of the second- through sixth-order results reveals a beautiful pattern of regularity [9]. From expressions (12.16) and (12.50), the sum of the second- through sixth-order amplitudes for quark-quark scattering asymptotically approaches

$$-\frac{sg^2}{2m^2} \frac{1}{\Delta^2} [1 - \bar{\alpha}(\Delta)\ln s + \tfrac{1}{2}\bar{\alpha}^2(\Delta)\ln^2 s] G_1$$

$$+ \frac{is}{4m^2} \left[g^4 I(\Delta) - \frac{3g^6}{2\pi} \ln s\, I_1(\Delta) \right] G_2$$

$$+ \frac{isg^6 \ln s}{8\pi m^2} [2I_1(\Delta) - \Delta^2 I^2(\Delta)] G_3 + \frac{sg^6}{12m^2} I_1(\Delta) G_4, \quad (12.51)$$

where the notation is explained in the discussion following expression (12.50). We observe the following.

1. The coefficient of G_1. As we have mentioned, the coefficient of G_1 is equal to the first three terms of the expansion of a Regge pole term. This Reggeization of the gluon can in fact be verified to all orders. The G_1 term in expression (12.51) is therefore the one-Reggeon exchange amplitude, the counterpart of the one-photon exchange amplitude in QED.

2. The coefficient of G_2. We mentioned in section 12.2 that the g^4 term in the coefficient of G_2 is the same as the two-vector-meson exchange amplitude in the Abelian gauge field theory. What is the meaning of the g^6 term in the coefficient of G_2? Because we have just seen that the propagator of a gluon of momentum $\mathbf{q}_\perp$ is modified by a factor of $s^{-\bar\alpha(\mathbf{q}_\perp)}$, the amplitude for the exchange of two Reggeized gluons must be modified accordingly. This modification is precisely the coefficient of G_2. To see this, we have from identity (12.3) that, as the gluons are Reggeized,

$$
I(\Delta) = \int \frac{d^2 q_\perp}{(2\pi)^2} \frac{1}{\mathbf{q}_\perp^2} \frac{1}{(\Delta - \mathbf{q}_\perp)^2}
$$

$$
\rightarrow \int \frac{d^2 q_\perp}{(2\pi)^2} \frac{\exp[-\bar\alpha(\mathbf{q}_\perp)\ln s]}{\mathbf{q}_\perp^2} \frac{\exp[-\bar\alpha(\Delta - \mathbf{q}_\perp)\ln s]}{(\Delta - \mathbf{q}_\perp)^2}.
$$

As we expand $\exp[-\bar\alpha(\mathbf{q}_\perp)\ln s]$ into $1 - \bar\alpha(\mathbf{q}_\perp)\ln s + \ldots$, the right-hand side of the above equation is

$$
I(\Delta) - \frac{3g^2 \ln s}{4\pi} \int \frac{d^2 q_\perp}{(2\pi)^2} \frac{1}{\mathbf{q}_\perp^2} \frac{1}{(\Delta - \mathbf{q}_\perp)^2} [\mathbf{q}_\perp^2 I(\mathbf{q}_\perp) + (\Delta - \mathbf{q}_\perp)^2 I(\Delta - \mathbf{q}_\perp)] + \ldots
$$

$$
= I(\Delta) - \frac{3g^2 \ln s}{2\pi} I_1(\Delta) + \ldots,
$$

where equations (12.17) and (10.49) have been used. This is precisely the quantity in the bracket of the coefficient of G_2 given by expression (12.51). Thus the G_2 term in expression (12.51) is the two-Reggeon exchange amplitude, the counterpart of the two-photon exchange amplitude in QED.

It is further observed that the G_2 term is related to the G_1 term by unitarity. The unitarity condition reads

$$
Im\,\mathcal{M}_{fi} = \frac{1}{2}\sum_n (2\pi)^4 \delta^{(4)}(P_i - P_n)\frac{\mathcal{M}_{fn}^* \mathcal{M}_{ni}}{\prod_m f_m}, \tag{12.52}
$$

where there is a kinematic factor f_m for each particle in the intermediate state n, with

$$
f = \begin{cases} E/m & \text{for a quark with energy } E \text{ and mass } m, \\ 2E & \text{for a gluon with energy } E. \end{cases}
$$

In the present consideration, f and i are states of two quarks. Let n be a two-quark state, and let both $\mathcal{M}_{fn}$ and $\mathcal{M}_{ni}$ be approximated by the G_1 term in expression (12.51); then the right-hand side of equation (12.52) is

$$\frac{1}{2}\int \frac{d^3q}{(2\pi)^3}(2\pi)\delta(2\sqrt{(\mathbf{p}_1+\mathbf{q})^2+m^2}-2\omega)\frac{m^2}{\omega^2}\left(\frac{sg^2}{2m^2}\right)^2 \frac{s^{-\bar{\alpha}\mathbf{q}_\perp}}{\mathbf{q}_\perp^2}\frac{s^{-\bar{\alpha}(\Delta-\mathbf{q}_\perp)}}{(\Delta-\mathbf{q}_\perp)^2}$$

$$\simeq \frac{sg^4}{4m^2}\int\frac{d^2q_\perp}{(2\pi)^2}\frac{s^{-\bar{\alpha}(\mathbf{q}_\perp)}}{\mathbf{q}_\perp^2}\frac{s^{-\bar{\alpha}(\Delta-\mathbf{q}_\perp)}}{(\Delta-\mathbf{q}_\perp)^2}, \qquad (12.53)$$

which is just the coefficient of G_2 in expression (12.51).

3. The coefficient of G_3. Just as the G_2 term is obtained from the elastic scattering amplitude of one-Reggeon exchange through the unitarity condition, the G_3 term is obtained from the inelastic amplitude of one-gluon production, illustrated in figure 12.11, through the unitarity condition. The calculation is straightforward and is not presented here. We observe that the G_3 term in expression (12.51) is the counterpart of the lowest-order tower amplitude discussed in the preceding chapter, with one gluon the counterpart of the fermion-antifermion pair. Higher-order terms Reggeize the virtual gluons in figure 12.11. There are also terms representing the contributions of the production of two or more gluons. These terms are the counterparts of the higher-order tower amplitudes. Explicit calculations have been carried out to the tenth order.

4. The coefficient of G_4. Just as the coefficients of G_1 and G_2 represent the amplitudes of one-Reggeon exchange and two-Reggeon exchange, respectively, the coefficient of G_4 represents the amplitude of three-Reggeon exchange. This is true because the coefficient of G_4 in the sixth order is precisely the sum of the space-time factors of all three-gluon exchange amplitudes. Also, the group factor G_4 is the group factor for the exchange of the three gluons, as can be seen from the definition of G_4 in figure 12.7.

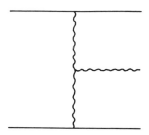

Figure 12.11
A diagram for the production of a gluon in quark-quark scattering.

The Reggeization of the gluons does not enter in the sixth order. It has been verified that the contributions of the eighth-order and the tenth-order perturbations give exactly this correction of the Reggeization.

Let us represent the coefficient of G_1 graphically by the diagram in figure 12.12a, where the exchanged particle, being a Regge pole, is represented by a shaded rectangle. Then the coefficient of G_2 is equal to the product of the two amplitudes represented by the two diagrams in figure 12.12b, with the cross implying the summation over the intermediate states. Furthermore, the coefficients of G_3 and G_4 are equal to the amplitude represented by figure 12.12c and that represented by figure 12.12d, respectively.

We recall that the sum of the multi-vector-meson exchange amplitudes in the Abelian gauge field theory is in the form of the eikonal formula

$$\frac{is}{2m^2} \int d^2 x_\perp \exp[i\Delta \cdot x_\perp (1 - e^{i\chi})]. \tag{12.54}$$

Indeed, if we make in expression (12.54) the expansion of

$$1 - e^{i\chi} = -i\chi + \frac{\chi^2}{2!} + i\frac{\chi^3}{3!} + \dots, \tag{12.55}$$

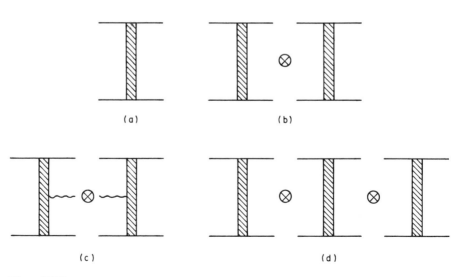

Figure 12.12
A diagrammatic representation for the coefficient of (a) G_1 (this coefficient is the amplitude for the exchange of a Regge pole), (b) G_2 (this coefficient is equal to the product of two elastic one-Reggeon-exchange amplitudes summed over all intermediate states), (c) G_3, and (d) G_4.

then the χ, χ^2, χ^3, ... terms give the amplitudes for the exchange of one vector meson, two vector mesons, three vector mesons, ..., respectively. Furthermore, this formula is in agreement with the result of the s-channel potential model discussed in chapter 4 and is explicitly unitary.

We have now identified that the counterparts in the non-Abelian gauge theory of the χ, χ^2, and χ^3 terms are precisely the G_1, G_2, and G_4 terms in expression (12.51) [9]. Indeed, in the non-Abelian case, the eikonal χ is given by the one-Reggeon-exchange amplitude (12.20) represented by the diagram in figure 12.13a. Thus the eikonal χ is s dependent. Furthermore, it has an additional group factor G_1. Hence the square of the eikonal has the group factor G_2, and the cube of the eikonal has the group factor G_4, in agreement with the diagrammatic calculation.

As we have mentioned, the G_3 term is the counterpart of the lowest-order tower amplitude discussed in chapter 11. The counterparts of the higher-order tower amplitudes have been explicitly calculated up to the tenth order. If we extrapolate these results to all orders, we can conclude that the sum of these terms violates the Froissart bound. More precisely, this sum has a branch cut in the complex angular momentum plane, with the end of the branch point located at [2, 4]

$$J = 1 + \frac{n \ln 2}{\pi^2} g^2 > 1, \tag{12.56}$$

for the case in which the symmetry group is $SU(n)$. This is the counterpart

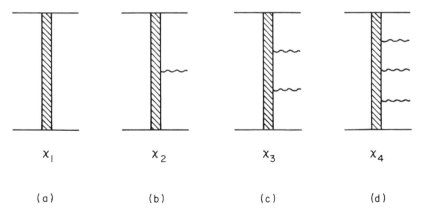

$$\chi_1 \qquad\qquad \chi_2 \qquad\qquad \chi_3 \qquad\qquad \chi_4$$

(a)   (b)   (c)   (d)

Figure 12.13
Diagrammatic representations of the terms in the eikonal.

of the result that the tower amplitude has a branch point in the angular momentum plane located at

$$J = 1 + \frac{11\pi}{32}\alpha^2 > 1,$$

violating the Froissart bound and unitarity.

How is unitarity restored? The form of the G_3 term, represented by figure 12.12c, suggests that the amplitude is given by the eikonal formula (12.54), with the eikonal χ having not only the term represented by the diagram in figure 12.13a but also the term represented by the diagram in figure 12.13b. Let us denote these terms by χ_1 and χ_2, respectively. Then the term χ_2^2 in the expansion of $\exp[i(\chi_1 + \chi_2)]$ has the group factor G_3 and also the correct coefficient. Indeed, it has been shown that all terms in the scattering amplitude up to the tenth perturbative order are given correctly by the eikonal formula with the eikonal equal to the sum of the four terms represented in figure 12.13. This is discussed more fully in the next chapter.

12.5 Scattering of Gluons

In the preceding sections we restricted ourselves to quark-quark scattering. If one or more of the incident particles are gluons, then the only modification of the space-time factor is an overall factor of $2m$ for each of the incident gluons [9]. For example, by using the Feynman rules in table 12.1, we find that the second-order gluon-gluon scattering amplitude corresponding to figure 12.14 asymptotically approaches

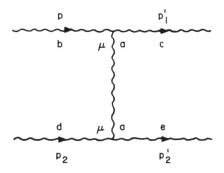

Figure 12.14
A second-order diagram for gluon-gluon scattering.

$$\frac{g^2}{t}[(p'_1 + p_1)_\mu if_{acb}][-(p'_2 + p_2)_\mu if_{ade}] = -\frac{2g^2 s}{\Delta^2} t_a^{(1)} t_a^{(2)}, \tag{12.57}$$

where $t_a^{(1)}$ (resp. $t_a^{(2)}$) is a matrix in the color space of the first (resp. second) gluon defined by

$$(t_a^{(1)})_{cb} = if_{acb} \tag{12.58}$$

and

$$(t_a^{(2)})_{ed} = -if_{ade}. \tag{12.59}$$

Note the difference of the signs in the right-hand sides of equations (12.58) and (12.59) in our convention, which is chosen so that a vertex on the top (resp. bottom) horizontal line contributes a space-time factor $2p_{1\mu}$ (resp. $2p_{2\mu}$). (If we had changed the sign in equation (12.59), the space-time factor for a vertex on the bottom horizontal line would have been $-2p_{2\mu}$.) Note that the space-time factor of equation (12.57) differs from that of expression (12.1) by a factor of $4m^2$. The same difference occurs in the higher-order diagrams. In addition, if we represent the group factors diagrammatically, then the diagrammatic equalities in figure 12.7 remain valid as we replace the quark lines at the top and the bottom by gluon lines.

One subtle point may be worth mentioning. For gluon-gluon scattering, some diagrams that appear to be different may be the same topologically. For example, the two diagrams in figure 12.15 are the same diagram. One might think that, as a consequence, the sixth-order quark-quark scattering amplitude, with contributions from both diagram 3 and diagram 4 in figure 12.7, differs qualitatively from the sixth-order gluon-gluon scattering amplitude. For both diagram 3 and diagram 4 of figure 12.7, however, the

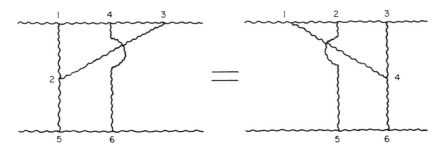

Figure 12.15
Two diagrams for gluon-gluon scattering that appear to be different but are actually the same.

large plus momentum of magnitude 2ω flows through the top quark line, whereas for the diagram in figure 12.15, the large plus momentum of magnitude 2ω may flow through either the route 1 4 3 or the route 1 2 3. The route 1 4 3 (resp. 1 2 3) of the diagram in figure 12.15, is the counterpart of the diagram 3 (resp. diagram 4) in figure 12.7. As a consequence, the coefficients of G_1, G_3, etc. for the gluon-gluon scattering amplitude are precisely $4m^2$ times those of the quark-quark scattering amplitude.

References

[1] B. M. McCoy and T. T. Wu, *Physical Review Letters* 35 (1975): 604, *Physical Review D* 13 (1976): 1076.

[2] V. S. Fadin, E. A. Kuraev, and L. N. Lipatov, *Physics Letters B* 60 (1975): 50.

[3] L. Łukaszuk and X. Y. Pham, *Physics Letters B* 53 (1974): 287; H. T. Nieh and Y. P. Yao, *Physical Review D* 13 (1976): 1082; L. Tyburski, *Physical Review D* 13 (1976): 1107; L. L. Frankfurt and V. E. Sherman, *Soviet Journal of Nuclear Physics* 23 (1976): 581.

[4] C. Y. Lo and H. Cheng, *Physical Review D* 13 (1976): 1131; P. S. Yeung, *Physical Review D* 13 (1976): 2306; H. Cheng and C. Y. Lo, *Physical Review D* 15 (1977): 2959.

[5] M. T. Grisaru, H. J. Schnitzer, and H. S. Tsao, *Physical Review Letters* 30 (1973): 811.

[6] F. Englert and R. Brout, *Physical Review Letters* 13 (1964): 321; P. W. Higgs, *Physics Letters* 12 (1964): 132, *Physical Review Letters* 13 (1964): 508, *Physical Review* 145 (1966): 1156; G. S. Guralnik, C. R. Hagen, and T. W. B. Kibble, *Physical Review Letters* 13 (1964): 585.

[7] For the Feynman rules in this theory, see G. 't Hooft and M. Veltman, in *Renormalization of Yang-Mills Fields and Applications to Particle Physics*, C. P. Korthals-Altes, ed. (Marseilles: CNRS, 1972).

[8] C. D. Stockham, *Physical Review D* 15 (1977): 1736; Ya. Ya. Balitskii and L. N. Lipatov, *Soviet Journal of Nuclear Physics* 28 (1978): 822, *JETP Letters* 30 (1979): 355.

[9] H. Cheng, J. A. Dickinson, and K. Olaussen, *Physical Review D* 23 (1981): 534.

13 Multiparticle Unitarity and the Extended Eikonal Formula

13.1 Introduction

In the two preceding chapters we made a detailed study of high-energy scattering both in the Abelian gauge field theory and in the non-Abelian gauge field theory. We have found a regular pattern that suggests that the high-energy scattering process is strongly absorptive. In chapter 7 we argued from physical grounds that this means that the total cross section increases indefinitely with energy.

The stage is now set for giving a comprehensive diagrammatic study to substantiate this conclusion. For this purpose it is helpful to have an overview of the role of each Feynman diagram in relation to the complete scattering amplitude. Let us begin with the simple case of the scattering of a nonrelativistic particle of momentum k in a potential $V(\mathbf{x})$. The optical theorem for the scattering amplitude f is

$$\mathrm{Im}\, f = \frac{k}{4\pi} \int |f|^2 \, d\Omega. \tag{13.1}$$

It is easy to see that the Born approximation of f does not satisfy equation (13.1). Indeed, let us approximate f in the right-hand side of equation (13.1) by the Born term $\tilde{V}$, which is proportional to the Fourier transform of V. Then equation (13.1) says that there must exist in the imaginary part of f a term that is quadratic in $\tilde{V}$. It then follows from the dispersion relation that the real part of f has a term quadratic in $\tilde{V}$. This means that, once we take into consideration the process of single scattering by the potential, we must, in order to satisfy unitarity and the dispersion relation, take into consideration the process of double scattering by the potential; however, the sum of the first Born term and the second Born term still does not satisfy equation (13.1), because the right-hand side of equation (13.1) contains in this approximation terms that are cubic and quartic in $\tilde{V}$. Thus we must bring in the third and the fourth Born terms. Doing this successively, we generate the whole Born series for f.

The role of each Born term in potential scattering has a quantum field theoretic analogue. For the fermion-fermion elastic scattering amplitude in the Abelian gauge field theory, the counterpart of the Born term $\tilde{V}$ is the lowest-order amplitude of one-vector-meson exchange. By the same reasoning (as will be outlined more precisely in the next section), once we take into account the diagram of one-vector-meson exchange, we must take into account the diagrams of two-vector-meson exchange. By iterating this further as before, we obtain all multi-vector-meson exchange diagrams. As we discussed in chapter 3, the sum of amplitudes of multi-vector-meson exchange is in the form of the eikonal formula (4.5), which is explicitly unitary. Furthermore, the eikonal δ in relation (4.5) is related to the one-vector-meson exchange amplitude. We refer to this procedure of iterating the one-vector-meson exchange amplitude as s-channel unitarization. Visually, the diagrams successively obtained from s-channel unitarization expand in the horizontal direction.

One of the major distinctions between the processes in a quantum field theory and those in potential scattering is that in a quantum field theory particles can be created or destroyed. As we discussed in chapter 6, these elastic and inelastic scattering amplitudes affect one another. For example, just as the existence of the one-vector-meson exchange diagram leads to that of the two-vector-meson exchange diagrams, the existence of the diagrams for pair creation illustrated in figure 5.2 leads to the tower diagrams illustrated in figure 6.1 for the elastic scattering. Similarly, the diagrams for the creation of two pairs lead to the tower diagrams illustrated in figure 6.2. More generally, the diagrams for n-pair creation lead to the tower diagrams of n loops. As we saw in chapter 6, the sum of these tower amplitudes is given by the contribution of a Regge branch cut (for other theories, a Regge pole). We refer to this procedure of taking care of the creation of more and more pairs as t-channel unitarization. Visually, the diagrams successively obtained from t-channel unitarization expand in the vertical direction.

Because the unitarity condition, the dispersion relation, and particle creation must be taken into account in any theory of high-energy scattering, both the s-channel unitarization and the t-channel unitarization should be incorporated. This is precisely what we will do in the next section.

13.2   Abelian Gauge Field Theory

The unitarity condition reads

$$\operatorname{Im} \mathscr{M}_{fi} = \tfrac{1}{2} \sum_n (2\pi)^4 \delta^{(4)}(\sum P_i - \sum P_n) \mathscr{M}_{fn}^* \mathscr{M}_{ni} / \prod_m f_m, \tag{13.2}$$

where a kinematic factor f_m is present for each particle in the intermediate state

$$f_n = \begin{cases} 2E_n & \text{for a boson,} \\ E_n/m & \text{for a fermion of mass } m. \end{cases} \tag{13.3}$$

Let both i and f in equation (13.2) be two-fermion states, and consider the contribution of a two-fermion intermediate state to the imaginary part of $\mathcal{M}_{fi}$ (that is, take n in the right-hand side of equation (13.2) to be a two-fermion state). If we replace $\mathcal{M}_{fn}$ and $\mathcal{M}_{ni}$ by the one-vector-meson exchange amplitude for elastic scattering, then the right-hand side of equation (13.2) yields the imaginary part of the amplitude corresponding to the box diagram of two-vector-meson exchange. The real part of the amplitude corresponding to the box diagram is then brought in by the dispersion relation. The requirement of gauge invariance brings in the crossed box diagram. Therefore the unitarity condition, the dispersion relation, and gauge invariance dictate that we must take into account the contribution of two-vector-meson exchange once we take into account one-vector-meson exchange. As we discussed in the preceding section, successive application of this argument brings in all diagrams of multi-vector-meson exchange and corresponds to the procedure of s-channel unitarization.

The procedure just outlined takes care of the *elastic* unitarity condition only. In other words, n in equation (13.2) has so far been restricted to a two-fermion state. As we have learned, *inelastic* scattering plays a decisive role in high-energy scattering. The lowest-order contributing diagrams for the reaction $f + f \to f + f + f + \bar{f}$ are the ones illustrated in figures 13.1b and 13.1c. They are the only fourth-order diagrams in which the fermion-

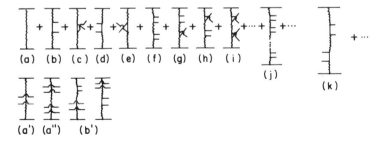

(a) (b) (c) (d) (e) (f) (g) (h) (i)   (j)   (k)

(a') (a'')   (b')

Figure 13.1
The lowest-order contributing diagrams for all reactions, elastic and inelastic. The s-channel is from left to right. The second row of diagrams is obtained from the first row of diagrams with the addition of one (a', b') and two (a'') free pairs. The longitudinal momenta of the particles in the chain are ordered.

antifermion pair can be created with small momenta in the center-of-mass system (that is, a pionization product). Again, we take f and i in equation (13.2) to be two-fermion states and replace $\mathcal{M}_{fn}$ and $\mathcal{M}_{ni}$ in equation (13.2) by the amplitudes corresponding to the diagrams in figures 13.1b and 13.1c. Then the term $\mathcal{M}_{fn}^* \mathcal{M}_{ni}$ in the right-hand side of equation (13.2) is equal to the imaginary part of the lowest-order tower amplitude, as we discussed in chapter 6. This is illustrated in figures 13.2a and 13.2b. The third tower diagram of one fermion loop, not illustrated in the figure, must be taken into account by the requirement of gauge invariance. Similarly, the lowest-order contributing diagrams for $f + f \to f + f + n$ pairs are the ones illustrated in figure 13.1j. By using the unitarity condition as before, we see that these diagrams lead to the n-pair one-tower diagrams. As we mentioned in the preceding section, summing over n corresponds to t-channel unitarization.

This diagrammatic demonstration of s-channel unitarization and t-channel unitarization also shows clearly why neither unitarization procedure is complete in itself. Each includes but a subset of diagrams. Furthermore, s-channel unitarization cannot take into account inelastic contributions, unless one puts in an absorptive "potential" by hand. On the other hand, t-channel unitarization gives an amplitude violating the Froissart bound. This is not surprising, as the resulting amplitude does not satisfy equation (13.2)—in the same way that the two-vector-meson exchange amplitude does not satisfy the elastic unitarity condition.

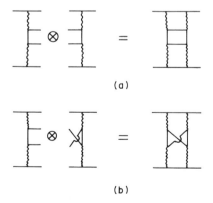

Figure 13.2
Schematic diagrams of elastic scattering generated by diagrams of inelastic scattering by means of the unitarity condition.

A scheme encompassing both s-channel unitarization and t-channel unitarization can be developed by following the guidance of the unitarity condition (13.2). We observe two important points.

1. The unitarity condition relates a Feynman diagram to others. Consequently, if we include one Feynman diagram, we must also include a set of related Feynman diagrams uniquely determined by the unitarity condition. Thus we are not allowed to take an arbitrary set of diagrams, say the tower diagrams, and expect the answer to be consistent with unitarity.

2. The unitarity condition couples all elastic and inelastic scattering amplitudes together. Consequently, a consistent scheme of constructing a unitary S-matrix must deal with the elastic and inelastic scattering amplitudes all at once and on the same footing.

For the sake of classification, we call the contributing lowest-order diagrams for each process the V diagrams. They are the counterparts of the potential V in nonrelativistic scattering. For example, the one-vector-meson exchange diagram for elastic scattering is a V diagram. For the reaction $f + f \rightarrow f + f + f + \bar{f}$, the contributing lowest-order diagrams are those illustrated in figures 13.1b and 13.1c (note that they are the one-vector-meson exchange diagrams for the aforementioned inelastic reaction) and are hence called V diagrams as well. Other V diagrams, all of them one-vector-meson exchange diagrams, are also illustrated in figure 13.1.

Let us substitute for the scattering amplitudes in the right-hand side of equation (13.2) the amplitudes corresponding to the V diagrams, elastic as well as inelastic. If f and i in equation (13.1) are both two-fermion states, we obtain, as mentioned before, the amplitude for the tower diagrams. We therefore call the tower diagrams the V^2 diagrams for elastic scattering.

It is important to keep in mind that, because we must treat inelastic scattering and elastic scattering on the same footing, we must extend the iteration procedure for the inelastic amplitudes as well. Let us therefore consider the case in which f is a state of two fermions and a fermion-antifermion pair, whereas i remains a two-fermion state. The amplitudes obtained by replacing $\mathcal{M}_{fn}$ and $\mathcal{M}_{ni}$ by the amplitudes of V diagrams correspond to the diagrams in figure 13.3. Note that these diagrams are the ones obtained from the tower diagrams by attaching a fermion-antifermion pair to one of the virtual vector mesons. More generally, for the reaction of $f + f \rightarrow f + f + n$ pairs, the resulting diagrams are those obtained from one-tower diagrams by attaching n pairs to the virtual vector meson lines. We call these diagrams the V^2 diagrams for n-pair creation.

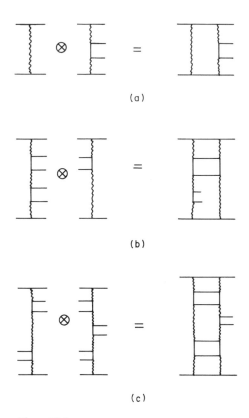

Figure 13.3
Schematic diagrams of inelastic scattering generated by lower-order diagrams by means of the unitarity condition.

The sum of V diagrams and V^2 diagrams still does not satisfy the unitarity condition (13.2). Indeed, let us now replace the amplitudes in the right-hand side of equation (13.2) by those corresponding to the V diagrams and V^2 diagrams. The diagrams obtained from this iteration are called V^3 diagrams and V^4 diagrams, respectively. Some examples of V^3 diagrams are illustrated in figure 13.4. Other examples of V^3 and V^4 diagrams are illustrated in the second and third rows of figure 13.5. By repeating this procedure, we can obtain in succession the $V^5 \ldots V^N \ldots$ diagrams for all reactions.

The diagrams generated by this procedure have a regular pattern. For the elastic reaction $f + f \to f + f$, the V^N diagrams can be obtained by the following steps: (1) Draw two horizontal lines representing the two

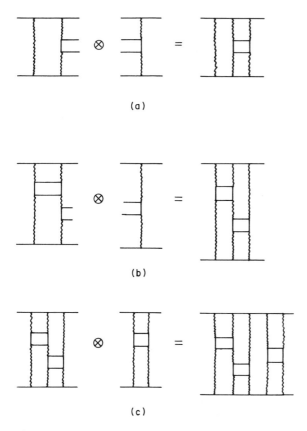

(a)

(b)

(c)

Figure 13.4
Schematic V^3 and V^4 diagrams generated by means of the unitarity condition.

incoming fermions; (2) from the top fermion line, draw downward N vector mesons; (3) choose any two of these N vector meson lines to be absorbed by a fermion loop, which emits two vector mesons to replenish the two absorbed; all orders of emission and absorption are allowed; (4) repeat step (3) until all N vector meson lines end at the bottom fermion line.

The V^N diagrams for the reaction $f + f \rightarrow f + f + n$ pairs are obtained from those for the elastic reaction by attaching n pairs to the virtual vector meson lines of the latter. All ways of attachment are allowed. We can similarly obtain the V^N diagrams for the reaction $f + f + n$ pairs $\rightarrow$ $f + f + n'$ pairs. Some examples are shown in the second and the third columns of figure 13.5.

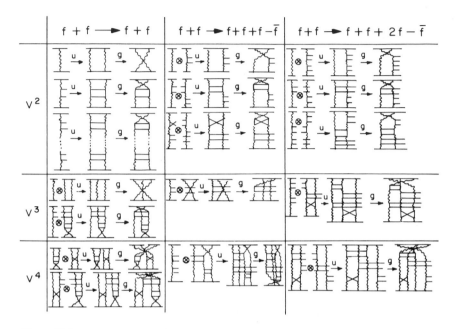

Figure 13.5
The generation of diagrams. The letters u and g denote unitarity and gauge invariance
respectively. All vertices are three-line vertices.

The diagrams obtained in this way incorporate both s-channel unitarization and t-channel unitarization. They extend in both the horizontal
direction and the vertical direction and represent both inelastic and elastic
reactions.

The V^2 terms for the elastic reaction $f + f \rightarrow f + f$ were calculated
in chapter 11. All other V^N terms, $N = 3, 4, \ldots$, for the elastic reaction
can be calculated in a straightforward way. We can also calculate all
V^N terms, $N = 1, 2, \ldots$, for the inelastic reaction $f + f \rightarrow f + f + n$ pairs.
This calculation [1] is tedious and is not given here, but the result is surprisingly simple. The S-matrix, expressed in the space of the impact distance b between the two incident fermions, is given by the extended eikonal
formula

$$S = e^{iV}, \tag{13.4}$$

where V is simply the operator with the matrix elements equal to the
asymptotic amplitudes of the diagrams in figure 13.1. To put it another

way, the scattering amplitude $\mathscr{M}_{fi}$ in the momentum space is given by

$$\mathscr{M}_{fi} = 2i(\textstyle\prod f_n)^{1/2} \int d^2 b \, e^{i\mathbf{\Delta}\cdot\mathbf{b}} (I - e^{iV})_{fi}, \tag{13.5}$$

where I is the unity matrix, $\mathbf{\Delta}$ is the momentum transfer received by one of the incident fermions, $\mathbf{b}$ is the impact distance between the two incident fermions, and $\prod f_n$ is the product of the kinematic factors of the external particles. Equation (13.5) has been verified to hold for all processes, inelastic as well as elastic. More precisely, if we expand e^{iV} in equation (13.5) into a Taylor series of V, then the term iV in the expansion, substituted into equation (13.5), gives precisely the asymptotic amplitudes for the V diagrams in figure 13.1. Furthermore, the term $(iV)^n$ in this expansion, substituted into equation (13.5), has been verified to give precisely the asymptotic amplitudes for the V^N diagrams for inelastic scattering as well as elastic scattering.

The operator V in equation (13.5) can be represented as

$$V = V_0 + V_1 + \ldots + V_n + \ldots . \tag{13.6}$$

In equation (13.6)

$$V_0 = -\frac{q^2}{2\pi} K_0(\lambda b), \tag{13.7}$$

and V_0 is contributed by figure 13.1a of elastic scattering. The operator V_1 involves the operators for creating a pair or for annihilating a pair in such a way that it produces the asymptotic amplitudes corresponding to figures 13.1b through 13.1e. Similarly for the operator V_n in equation (13.6).

Both the eikonal model and the Regge model can be seen as approximations of equation (13.5). If we approximate V by V_0, then equation (13.5) becomes the eikonal formula (3.36)—the equation on which models of s-channel unitarization are based. If we expand e^{iV} in equation (13.5) into a Taylor series of V, then the V^2 term gives the asymptotic amplitudes corresponding to the V^2 diagrams. In particular, it gives for the elastic scattering amplitude the sum of the tower amplitudes, which has Regge (pole or cut) behavior usually obtained in models of t-channel unitarization. If we define the V^2 term as the "bare Pomeron," then the V^4, V^6, ... terms can be identified as the two-, three-, ... Pomeron exchange. It is, however, not fruitful to expand the amplitude in this way, as the resulting series is divergent. Equation (13.5) can be regarded as a generalized eikonal formula, with the eikonal V an operator. Because V is a Hermitian operator, this elegant eikonal formula is manifestly unitary.

13.3   Non-Abelian Gauge Field Theory

In the preceding section, we showed that the high-energy scattering amplitudes in the Abelian gauge field theory are given by the eikonal form e^{iV}, where V is related to the sum of the lowest-order amplitudes for the elastic and the inelastic reactions. In this section we turn our attention to the high-energy scattering amplitudes in the non-Abelian gauge field theory.

The non-Abelian gauge field theory is different from the Abelian gauge field theory in many aspects. Consequently, what holds in the Abelian gauge field theory may not hold in the non-Abelian gauge field theory. Consider, for example, the physical meaning of exponentiation in the Abelian gauge field theory. The first term in V, called V_0, is related to the Yukawa potential between two fermions, and the rest of the terms represent inelastic contributions. Thus we can think of V as the generalized static field (elastic and inelastic) set up by the target fermion and e^{iV} as the amplitude of the scattering of the projectile fermion from this potential. This interpretation is possible because the target fermion remains a fermion after emitting a vector meson. In contrast, the gluon in the non-Abelian gauge theory carries color. The color state of the target particle therefore changes as it interacts with the projectile particle, and the concept of a static field does not apply. In other words, the field set up by the target particle depends on the scattering process that we seek to describe. Therefore we cannot expect to obtain the correct answer by exponentiating the lowest-order amplitudes, although such an expression is always unitary. Rather, we must find the answer by calculation.

How do the high-energy amplitudes in the Yang-Mills theory achieve unitarity? Calculations show [2, 3] that this is done in a most beautiful manner. Simply, the scattering amplitudes again add up to an eikonal form; the eikonal, however, is equal to an operator representing the sum of the lowest-order amplitudes of all reactions, with the *gluon propagators replaced by Regge poles*. Thus the counterpart to V_0, for example, is not the exchange of a gluon but that of a Regge pole on which the gluon lies. The energy of the projectile, therefore, enters the expression; that is, the eikonal is dependent on both the target and the projectile.

Let us give a brief summary of the results of calculations. For the elastic quark-quark scattering, it has been shown [4] that the sum of the leading terms is equal to

$$-\frac{g^2}{2m^2} \frac{s^{\alpha(\Delta)}}{\Delta^2 + \lambda^2} G_1,$$

(13.8)

where $\alpha(\Delta)$ is given by identity (12.21) and G_1 is the first group factor illustrated in the first column of figure 12.7. We note that, if we replace $\alpha(\Delta)$ by unity, expression (13.8) becomes the lowest-order amplitude of one-gluon exchange. Similarly, the sum of leading terms of the one-gluon creation amplitude for quark-quark scattering is obtained from the lowest-order amplitude by modifying each of the gluon propagators by a factor $s^{\alpha(\mathbf{q})}$, where $\mathbf{q}$ is the momentum of the gluon in question [5].

We showed in chapter 12 that up to the sixth perturbative order the eikonal formula (13.5) holds for quark-quark scattering with V replaced by χ, where

$$\chi = \sum_{n=1}^{\infty} \chi_n,$$

(13.9)

with the χ_n represented diagrammatically in figure 12.13.

This has actually been verified up to the tenth perturbative order [3]. Unlike the Abelian case, it is not possible in the non-Abelian case to single out a set of diagrams for summation, as all diagrams of a given perturbative order are related by gauge invariance. Therefore it is difficult to prove the validity of the eikonal formula to all perturbative orders. The verification of the eighth order and the tenth order is, however, highly nontrivial, requiring twenty-six coefficients extracted from the Feynman amplitudes to agree with the corresponding ones extracted from the eikonal formula. The calculations are not reproduced here. The interested reader is referred to [3] where the details and the nature of the approximations are discussed and where some useful methods for calculating the asymptotic amplitudes in non-Abelian gauge field theories are outlined.

13.4 Consequences of the Extended Eikonal Formula

We have found that in gauge field theories the S-matrix operator at high energies is given by

$$S(\mathbf{b}, T) = e^{i\chi(\mathbf{b}, T)}.$$

(13.10)

In equation (13.10) **b** is the impact distance between the two incident particles and

$$T \equiv \frac{1}{2\pi} \ln s.$$

The eikonal χ is equal to V of equation (13.6) in the Abelian case and is given by equation (13.9) in the non-Abelian case. Unlike the eikonal formula in potential scattering, to which it bears a formal resemblance, equation (13.10) is not yet in a form from which physical consequences can be readily extracted. Because the eikonal χ describes creation and annihilation of particles occurring in hadron scattering, it is an operator. As a result, the matrix elements of $e^{i\chi}$ are not related to those of χ in a simple manner, and the physical implications of the eikonal formula (13.10) remain to be deduced.

The expressions for the eikonal χ in gauge field theories are fairly complicated. In the Abelian case the complication is mostly due to the fact that a created fermion must be accompanied by a created antifermion, whereas in the non-Abelian case the complication is due to the complexity of the vertex factors and the Reggeization of the propagators. As a first attempt to understand the consequences of eikonalization, we consider a simplified model [6] in which a particle can be created singly with the vertex factor equal to the coupling constant g and the propagator in the momentum space is simply $(q_\perp^2 + \lambda^2)^{-1}$, where $q_\perp$ and λ are the transverse momentum and the mass of the virtual particle, respectively. For this model we have

$$\chi(\mathbf{b}, T) = g^2 \left[K(b) + g \int d^2 b_1 \int_0^T dT_1 \, K(|\mathbf{b} - \mathbf{b}_1|) x(\mathbf{b}_1, T_1) K(b_1) + \dots \right.$$

$$+ g^n \int \prod_{i=1}^n d^2 b_i \int_0^T dT_1 \int_0^{T_1} dT_2 \dots \int_0^{T_{n-1}} dT_n$$

$$\times K(|\mathbf{b} - \mathbf{b}_1|) x(\mathbf{b}_1, T_1) K(|\mathbf{b}_1 - \mathbf{b}_2|) x(\mathbf{b}_2, T_2) \dots K(b_n) + \dots]. \quad (13.11)$$

In equation (13.11), b equals $|\mathbf{b}|$, $K(b) = K_0(\lambda b)/2\pi$ is the Fourier transform of $(q_\perp^2 + \lambda^2)^{-1}$, with K_0 the modified Hankel function, and

$$x(\mathbf{b}_i, T_i) = [a^\dagger(\mathbf{b}_i, T_i) + a(\mathbf{b}_i, T_i)]/\sqrt{2},$$

with $a^\dagger(\mathbf{b}_i, T_i)$ and $a(\mathbf{b}_i, T_i)$ the operators for creating and annihilating, respectively, a particle at $\mathbf{b}_i$ with rapidity $2\pi T_i$ (in the laboratory system). The creation and annihilation operators satisfy the commutation rule

$[a(\mathbf{b}_i, T_i), a^\dagger(\mathbf{b}_j, T_j)] = \delta(T_i - T_j)\delta^{(2)}(\mathbf{b}_i - \mathbf{b}_j).$

Thus the first term in equation (13.11) represents the elastic contribution, the second term the contribution from the creation or annihilation of a single particle, and so on.

As an artifice to facilitate the calculation of $S(\mathbf{b}, T)$, we divide the $\mathbf{b}$- and T-spaces into small regions and approximate the integrals in equation (13.11) by sums. (Eventually, we recover these integrals by going to the limit when these small regions are infinitesimal.) Thus we replace the $\mathbf{b}$-space by a two-dimensional lattice with the lattice constant d (distance between two neighboring lattice points) and the T-space by a one-dimensional lattice with the lattice constant ε. We use the index j (two components implied) to denote a lattice point in the $\mathbf{b}$-space, and the index n to denote a lattice point in the T-space. We also replace the creation and annihilation operators by $a_j^\dagger(n)$ and $a_j(n)$, which satisfy the commutation rule

$[a_i(n), a_j^\dagger(m)] = \delta_{ij}\delta_{nm}.$

With such replacements and some algebraic manipulation, the eikonal $\chi(\mathbf{b}, T)$ in (13.11), denoted in the discretized version by $\chi_j(N)$, can be shown to be equal to the jth component of a vector $\chi(N)$, where

$$\chi(N) = [I + 2\sqrt{\varepsilon}\Lambda x(N)][I + 2\sqrt{\varepsilon}\Lambda x(N-1)]\ldots[I + 2\sqrt{\varepsilon}\Lambda x(1)]\chi(0).$$
$$(13.12)$$

In equation (13.12), $N\varepsilon = T$, I is the identity matrix, and $\chi(N)$ is a vector that has a component associated with each lattice point in the $\mathbf{b}$-space, with $(\chi(0))_j = g^2 K(|\mathbf{b}_j|)$. Also, Λ and $x(n)$ in equation (13.12) are matrices with the matrix elements

$$\Lambda_{ij} = dgK(|\mathbf{b}_i - \mathbf{b}_j|)/2, \qquad [x(n)]_{ij} = x_i(n)\delta_{ij},$$

associated with each pair of lattice points i and j, where

$$x_i(n) = [a_i(n) + a_i^\dagger(n)]/\sqrt{2}.$$

Because the operators $x_j(n)$ commute with one another, an eigenstate of $\chi_i(N)$ is a product of the eigenstates of the operators $x_j(n)$ involved in equation (13.12). In other words, each eigenstate of $\chi_i(N)$ is specified by a designation of the quantum numbers of $x_j(n)$ for all j and for all positive $n \leqslant N$.

It is therefore possible to calculate $S(\mathbf{b}, T)$ in equation (13.10), denoted by $S_j(N)$ in the discretized version, by expanding the ground state $|0\rangle$

into a superposition of the eigenstates of $\chi_j(N)$. Because the ground-state wave function of a harmonic oscillator is $\exp(-\frac{1}{2}x^2)/\pi^{1/4}$, we have from equations (13.10) and (13.12) that

$$S_j(N) = \int \prod_{1 \leqslant n \leqslant N} \prod_j \{dx_j(n)\exp[-x_j^2(n)]/\sqrt{\pi}\}$$

$$\times \exp(i\{[I + 2\sqrt{\varepsilon}\Lambda x(N)][I + 2\sqrt{\varepsilon}\Lambda x(N-1)]\ldots[I + 2\sqrt{\varepsilon}\Lambda x(1)]\chi(0)\}_j).$$

$$\tag{13.13}$$

We can think of $x(n)$ in equation (13.13) as a diagonal matrix whose matrix elements $x_j(n)$ are random variables with Gaussian distributions. Thus $S_j(N)$ is equal to the expectation value of the exponential of the jth component of a random vector $i\chi(N)$, which by equation (13.12) is equal to a product of random matrices operating on $i\chi(0)$. We study $S_j(N)$ in the limit $d \to 0$, $\varepsilon \to 0$, with $T = \varepsilon N \gg 1$.

The physical meaning of the eikonal form (13.13) is suggestive. Let us imagine a three-dimensional lattice with its lattice points specified by the index (j, n). We can think of j as the index specifying the transverse position on the lattice and n as the index specifying the longitudinal position on the lattice. Although the transverse dimension of the lattice is infinite, the longitudinal dimension of the lattice is equal to $N\varepsilon = T$. As $s = e^{2\pi T}$ increases, the longitudinal dimension of the lattice also increases.

There is, associated with each lattice point (j, n), a harmonic oscillator with the creation operator $a_j^\dagger(n)$ and the annihilation operator $a_j(n)$. When two high-energy particles collide, they can excite any of these harmonic oscillators associated with the three-dimensional lattice in any arbitrary manner. The scattering is therefore a stochastic process in which quanta of the harmonic oscillators are created and annihilated in a random way. It is interesting to observe that the relevant physical entity that directly enters is not the creation operator or the annihilation operator separately but the combination $x = (a + a^\dagger)/\sqrt{2}$. The eigenvalues of x play the role of a random variable that can take any value between $-\infty$ and ∞, with the probability distribution equal to the Gaussian $e^{-x^2}/\sqrt{\pi}$. It is also important to observe that the random variables $x_j(n)$ enter in the form of a power series for the eikonal χ, not for the S-matrix. As s becomes larger and larger, the three-dimensional lattice expands in the longitudinal direction, and more and more harmonic oscillators are involved. Thus χ receives contributions from an increasing number of random variables as s increases. Consequently, the expectation value of $\langle 0 | \chi_j^2(N) | 0 \rangle$, for example, is very large as $s \to \infty$. Indeed, this expectation value corresponds to the sum of tower diagrams and violates the Froissart bound. It is important that the S-matrix $S_j(N)$, being equal to $\langle 0 | e^{-\chi_j(N)} | 0 \rangle$, always satisfy the unitarity conditions, no matter how large $\chi_j(N)$ becomes. Indeed, let $\rho_{j,N}(\chi)$ be the probability that the eigenvalue of $\chi_j(N)$ is equal to χ; then we have

$$S_j(N) = \int_{-\infty}^{\infty} d\chi \, \rho_{j,N}(\chi) e^{i\chi}.$$

If $\rho_{j,N}(\chi)$ is concentrated in the region where χ is large, we expect that the rapid oscillation of the integrand makes the integral vanish.

We show in what follows that this is indeed the case. Let us define ξ as a vector that has a component associated with each lattice point of the **b**-space, and

$$S(\xi, N\varepsilon) \equiv \int \prod_{n=1}^{N} \prod_j \{dx_j(n) \exp[-x_j^2(n)]/\sqrt{\pi}\} \exp[i\chi(N)\cdot\xi]. \tag{13.14}$$

If we set $\xi_j = 1$ for a certain j and set all other components of ξ to zero, then $S(\xi, N\varepsilon)$ is equal to $S_j(N)$ given by equation (13.13). Thus a knowledge of $S(\xi, T)$ contains more than the complete information for the S-matrix over the whole **b**-lattice.

Equation (13.12) gives the recursion formula

$$\chi(N + 1) = [I + 2\sqrt{\varepsilon}\Lambda x(N + 1)]\chi(N).$$

By making use of this and taking the desired limit $\varepsilon \to 0$, it is straightforward to show that

$$\frac{\partial S(\xi, T)}{\partial T} = HS(\xi, T), \tag{13.15}$$

where

$$H = \sum_j \left(\sum_i \Lambda_{i-j}\xi_i \right)^2 \frac{\partial^2}{\partial \xi_j^2} \tag{13.16}$$

with the initial condition $S(\xi, 0) = \exp[i\xi \cdot \chi(0)]$.

A standard way to analyze the partial differential equation (13.15) is to perform a Laplace transform with respect to T. Let us define

$$\tilde{S}(\xi, w) \equiv \int_0^{\infty} dT \, e^{-wT} S(\xi, T). \tag{13.17}$$

Then

$$S(\xi, T) = \int_{\delta-i\infty}^{\delta+i\infty} \frac{dw}{2\pi i} e^{wT} \tilde{S}(\xi, w),$$ (13.18)

where δ is any positive number. Then

$$\tilde{S}(\xi, w) = -(H - w)^{-1} \exp[i\xi \cdot \chi(0)].$$

It is possible to prove that, in the desired limit $d \to 0$, $\tilde{S}(\xi, w)$ is an entire function of w. Thus we can set $\delta = 0$ in equation (13.18), and w becomes purely imaginary on the entire contour of integration. Consequently, as $T \to \infty$, the integrand in equation (13.18) oscillates rapidly, and $S(\xi, T)$ vanishes as $T \to \infty$ with ξ fixed. This means that the opacity of a particle at a fixed impact distance b, given by $1 - S(b, T)$, goes to unity as $T \to \infty$; that is, a particle becomes entirely absorptive at infinite energy. The reader is referred to reference [7] for this proof as well as for a more detailed discussion of the properties of the extended eikonal formula.

In summary, a systematic diagrammatic study led to the extended eikonal formula explicitly given by equation (13.4) in the Abelian gauge field theory and by equation (13.10) in the non-Abelian gauge field theory. This formula yields not only the asymptotic elastic scattering amplitude but also the asymptotic inelastic scattering amplitude. It contains terms in the Regge model as well as terms in the eikonal model and is manifestly unitary. One of its consequences is that a particle becomes more and more absorptive as energy increases and that the total cross section rises indefinitely with energy.

References

[1] H. Cheng, J. A. Dickinson, C. Y. Lo, K. Olaussen, and P. S. Yeung, *Physics Letters B* 76 (1978):129.

[2] H. Cheng, J. A. Dickinson, C. Y. Lo, and K. Olaussen, *Lettere al Nuovo Cimento* 25 (1979):175.

[3] H. Cheng, J. A. Dickinson, and K. Olaussen, *Physical Review D* 23 (1981):534.

[4] See references [1–5] of chapter 12.

[5] J. A. Dickinson, *Physical Review D* 16 (1977):1863.

[6] This model is equivalent to one treated by R. Blankenbecler and H. M. Fried, *Physical Review D* 8 (1973):678. For other references on the eikonal model, see G. Calucci, R. Jengo, and C. Rebbi, *Nuovo Cimento A* 4 (1971):330, 6 (1971):601; R. Aviv, R. L. Sugar, and R. Blankenbecler, *Physical Review D* 5 (1972):3252; S. Auerbach, R. Aviv, R. L. Sugar, and R. Blankenbecler, *Physical Review D* 6 (1972):2216.

[7] H. Cheng, J. A. Dickinson, P. S. Yeung, and K. Olaussen, *Physical Review D* 23 (1981):1411, *Physical Review Letters* 40 (1978):1681.

Appendix A

In this appendix we present the calculation of the vector meson impact factor $\mathscr{I}^V$ given by equations (9.36) through (9.38) [1].

We use in the following calculations the notation introduced in identities (11.2):

$$k_1 \equiv r_2 - r_1, \qquad k_2 \equiv r_2 + r_1,$$

$$p_1 \equiv r_3 + r_1, \qquad p_2 \equiv r_3 - r_1. \qquad (A1)$$

Thus

$$\Delta = 2r_1. \qquad (A2)$$

We work in the center-of-mass frame. We choose the z-axis to be in the direction of $\mathbf{r}_2$ and the x-axis to be in the direction of $\mathbf{r}_1$. Note that this is possible because

$$r_1 r_2 = -\mathbf{r}_1 \cdot \mathbf{r}_2 = 0.$$

Thus we have explicitly

$$r_1 = [0, \ 0, \ |\mathbf{r}_1|, \ 0],$$

$$r_2 = [\omega, \ \sqrt{\omega^2 - \mathbf{r}_1^2 - \lambda^2}, \ 0, \ 0],$$

$$r_3 = [\omega, \ -\sqrt{\omega^2 - \mathbf{r}_1^2 - m^2}, \ 0, \ 0], \qquad (A3)$$

where we have used the notation $[P_0, P_z, P_x, P_y]$. The three relevant sixth-order Feynman diagrams for $f + V \rightarrow f + V$ are redrawn in figure A.1 in this notation.

Let us first discuss briefly the polarizations of a massive vector meson. For a vector meson of four-momentum k, its polarization vector ε is a four-vector that satisfies

$$\varepsilon k = 0, \qquad \varepsilon^2 = -1. \qquad (A4)$$

If the vector meson has a nonzero mass, then there are three independent polarization vectors that satisfy equations (A4). The usual choice, which we follow, is that two are transverse (that is, $\varepsilon \cdot \mathbf{k} = 0$) and one is longitudinal (that is, $\varepsilon \| k$).

In chapter 9 we restricted the discussions to transverse polarization vectors only. In this appendix we discuss also the longitudinal case.

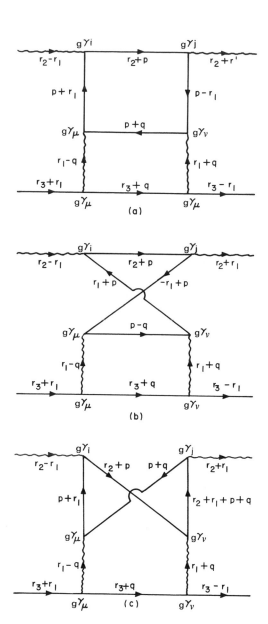

Figure A.1
Dominant sixth-order diagrams for fermion–vector meson scattering.

For the incoming photon of momentum $r_2 - r_1$, the two transverse polarization vectors and the longitudinal polarization vector are for high energies

$$[0, \ 0, \ 0, \ 1],\tag{A5}$$

$$[0, \ |\mathbf{r}_1|/\omega, \ 1, \ 0],\tag{A6}$$

$$\left[\omega - \frac{\lambda^2}{2\omega}, \ \omega - \frac{\mathbf{r}_1^2}{2\omega}, \ -|\mathbf{r}_1|, \ 0\right]\bigg/\lambda,\tag{A7}$$

respectively. Some components of the longitudinal polarization vector are as large as ω/λ as ω becomes large and give rise to large terms that cancel among themselves. In order to avoid complicated calculations, it is helpful to invoke gauge invariance and to subtract k_1/λ from expression (A7). Then the longitudinal polarization vector becomes

$$\left[-\frac{\lambda}{2\omega}, \ \frac{\lambda}{2\omega}, \ 0, \ 0\right].\tag{A8}$$

Similarly, for the outgoing photon of momentum $r_2 + r_1$, the two transverse polarization vectors and the longitudinal polarization vector are

$$[0, \ 0, \ 0, \ 1],\tag{A9}$$

$$[0, \ -|\mathbf{r}_1|/\omega, \ 1, \ 0],\tag{A10}$$

$$\left[-\frac{\lambda}{2\omega}, \ \frac{\lambda}{2\omega}, \ 0, \ 0\right],\tag{A11}$$

respectively.

Furthermore, it is useful to observe that, as a consequence of equations (A3) and (A4), we have for the transverse polarization vectors

$$r_{2i} = r_{1i} \quad \text{and} \quad r_{2j} = -r_{1j},\tag{A12}$$

where i and j denote the polarization directions of the incoming and the outgoing photons, respectively. For the longitudinal polarization vector ε given by expression (A8) or (A11),

$$\varepsilon r_2 = -\lambda, \qquad \varepsilon r_1 = 0.\tag{A13}$$

From equation (9.37), the function $\mathscr{I}_1^V$ in the present notation is

$$\mathscr{I}_1^V(\mathbf{r}_1, \mathbf{q}_\perp) = \frac{1}{4} g^4 \omega^{-1} \int_{-\infty}^{\infty} \frac{dq_-}{2\pi} \int \frac{d^4 p}{(2\pi)^4}$$

$$\times \frac{1}{[(r_2 + p)^2 - m^2][(p + r_1)^2 - m^2][(p - r_1)^2 - m^2]}$$

$$\times \operatorname{Tr} \left\{ \not{\varepsilon}_i (\not{p} + \not{r}_1 + m) \left[\frac{(\gamma_0 + \gamma_3)(\not{p} + \not{q} + m)(\gamma_0 + \gamma_3)}{(p + q)^2 - m^2 + i\varepsilon} \right. \right.$$

$$\left. + \frac{(\gamma_0 + \gamma_3)(\not{p} - \not{q} + m)(\gamma_0 + \gamma_3)}{(p - q)^2 - m^2 + i\varepsilon} \right] (\not{p} - \not{r}_1 + m) \not{\varepsilon}_f (\not{r}_2 + \not{p} + m) \right\}, \quad (A14)$$

where we have put

$$\not{r}_3 \sim \omega(\gamma_0 + \gamma_3) \tag{A15}$$

and where q_+ is understood to be zero. In equation (A14), ε_i and ε_f are the polarization vectors of the incoming and the outgoing vector mesons, respectively. The momenta of the two exchanged vector mesons are denoted by $r_1 - q$ and $r_1 + q$, respectively, instead of q and $\Delta - q$. We use the same notation for the impact factor.

We make the following observations.

1. $\not{p}$ can be written as

$$\not{p} = \tfrac{1}{2} p_+ (\gamma_0 + \gamma_3) + \tfrac{1}{2} p_- (\gamma_0 + \gamma_3) - \mathbf{p}_\perp \cdot \boldsymbol{\gamma}_\perp. \tag{A16}$$

Because

$$(\gamma_0 + \gamma_3)^2 = 0, \tag{A17}$$

the p_- in equation (A14) can be ignored if $\not{p}$ is adjacent to $(\gamma_0 + \gamma_3)$.

2. We have

$$(r_2 + p)^2 - m^2 + i\varepsilon = (r_2 + p)_+ (r_2 + p)_- - \mathbf{p}_\perp^2 - m^2 + i\varepsilon$$

$$\sim 2\omega\beta \left(\frac{\lambda^2 + \mathbf{r}_1^2}{2\omega} + p_- \right) - \mathbf{p}_\perp^2 - m^2 + i\varepsilon, \tag{A18}$$

where we have used equations (A3) and the notation

$$p_+ = -2\omega(1 - \beta). \tag{A19}$$

Similarly, we have

$$(p \pm r_1)^2 - m^2 + i\varepsilon = -2\omega(1 - \beta)p_- - (\mathbf{p}_\perp \pm \mathbf{r}_1)^2 - m^2 + i\varepsilon \tag{A20}$$

and

$$(p + q)^2 - m^2 + i\varepsilon = -2\omega(1 - \beta)(p_- + q_-) - (\mathbf{p}_\perp + \mathbf{q}_\perp)^2 - m^2 + i\varepsilon. \tag{A21}$$

Imagine now that we carry out the integration over p_-. We see that unless

$$0 < \beta < 1, \tag{A22}$$

all the singularities of the integrand in equation (A14) in the complex p_--plane are located on the same side of the real axis. By observation 1, the integrand in equation (A14) vanishes as p_-^{-3} at the infinity of the p_--plane. Thus, unless inequality (A22) is satisfied, the integration over p_- gives zero. We therefore restrict β to the region (A22).

3. Because of equation (A17), we have

$$(\gamma_0 + \gamma_3)(\not{p} \pm \not{q} + m)(\gamma_0 + \gamma_3) = 2p_+(\gamma_0 + \gamma_3) = -4(1 - \beta)\omega(\gamma_0 + \gamma_3), \tag{A23}$$

and

$$\frac{(\gamma_0 + \gamma_3)(\not{p} + \not{q} + m)(\gamma_0 + \gamma_3)}{(p + q)^2 - m^2 + i\varepsilon} + \frac{(\gamma_0 + \gamma_3)(\not{p} - \not{q} + m)(\gamma_0 + \gamma_3)}{(p - q)^2 - m^2 + i\varepsilon}$$

$$\sim 4\pi i(\gamma_0 + \gamma_3)\delta(q_-). \tag{A24}$$

By relation (A24), equation (A14) is immediately reduced to

$$\mathcal{I}_1^V(\mathbf{r}_1, \mathbf{q}_\perp) = \frac{1}{2}ig^4 \int \frac{d^2 p_\perp}{(2\pi)^2} \int_0^1 \frac{d\beta}{2\pi} \int_{-\infty}^\infty \frac{dp_-}{2\pi}$$

$$\times \frac{\mathrm{Tr}[\not{\epsilon}_i(\not{p} + \not{r}_1 + m)(\gamma_0 + \gamma_3)(\not{p} - \not{r}_1 + m)\not{\epsilon}_f(\not{r}_2 + \not{p} + m)]}{[(r_2 + p)^2 - m^2][(p + r_1)^2 - m^2][(p - r_1)^2 - m^2]}. \tag{A25}$$

Next we carry out the integration over p_- by closing the contour in the lower half plane. In this half plane the only singularity of the integrand is a pole located at the zero of the right-hand side of relation (A18). At this pole

$$p_- \sim \frac{\mathbf{p}_\perp^2 + m^2 - \beta(\lambda^2 + \mathbf{r}_1^2)}{2\omega\beta}, \tag{A26}$$

and

$$(p \pm r_1)^2 \sim -\frac{(\mathbf{p}_\perp \pm \beta \mathbf{r}_1)^2 + (1 - \beta)m^2 - \beta(1 - \beta)\lambda^2}{\beta}. \tag{A27}$$

Thus equation (A25) becomes

$$\mathscr{I}_1^V(\mathbf{r}_1, \mathbf{q}_\perp) = \frac{1}{8\pi} g^4 \int \frac{d^2 p_\perp}{(2\pi)^2} \int_0^1 d\beta$$

$$\times \frac{N_1}{[(\mathbf{p} + \beta \mathbf{r}_1)^2 + m^2 - \beta(1 - \beta)\lambda^2][(\mathbf{p}_\perp - \beta \mathbf{r}_1)^2 + m^2 - \beta(1 - \beta)\lambda^2]}, \tag{A28}$$

where

$$N_1 = \omega^{-1} \beta \operatorname{Tr}[\not\epsilon_i(\not p + \not r_1 + m)(\gamma_0 + \gamma_3)(\not p - \not r_1 + m)\not\epsilon_f(\not r_2 + \not p + m)], \tag{A29}$$

with p_+ and p_- given by relations (A19) and (A26), respectively.

It remains to evaluate N_1. If both ε_i and ε_f are transverse, we replace $\not\epsilon_i$ by γ_i and $\not\epsilon_f$ by γ_j. Thus we have

$$N_1 \sim -2\beta(1 - \beta) \operatorname{Tr}[\gamma_i(\not p - \not r_1 + m)\gamma_j(\not r_2 + \not p + m)]$$

$$- 2\beta(1 - \beta) \operatorname{Tr}[\gamma_i(\not p + \not r_1 + m)\gamma_j(\not r_2 + \not p + m)]$$

$$+ 2\beta^2 \operatorname{Tr}[\gamma_i(\not p + \not r_1 + m)(\not p - \not r_1 - m)\gamma_j]$$

$$= -16\beta(1 - \beta)\{p_i(r_2 + p)_j + (r_2 + p)_i p_j + \delta_{ij}[p(r_2 + p) - m^2]\}$$

$$+ 8\beta^2[2r_{1i}p_j - 2p_i r_{1j} - \delta_{ij}(p^2 - r_1^2 - m^2)]. \tag{A30}$$

From equation (A19), we have

$$p_i \sim -(1 - \beta)r_{2i} + p_{\perp i} = -(1 - \beta)r_{1i} + p_{\perp i}, \tag{A31}$$

where equations (A12) have been used. Similarly,

$$p_j \sim (1 - \beta)r_{1j} + p_{\perp j}. \tag{A32}$$

Thus we have, when ε_i and ε_f are both transverse,

$$N_1 \sim 32\beta(1 - \beta)(\beta^2 r_{1i} r_{1j} - p_{\perp i} p_{\perp j}) + 8\mathbf{p}_\perp^2 - 8\beta^2 \mathbf{r}_1^2 + 8m^2, \tag{A33}$$

where terms linear in $\mathbf{p}_\perp$ have been dropped, as these terms are integrated to zero when substituted into equation (A28).

When both ε_i and ε_f are longitudinal, we have after some calculations

$$N_1 \sim 32\lambda^2 \beta^2 (1 - \beta)^2. \tag{A34}$$

When ε_i is transverse and ε_f is longitudinal, we have after some calculations

$$N_1 \sim -16\beta^2(1-\beta)(1-2\beta)\lambda(\mathbf{r}_1 \cdot \boldsymbol{\varepsilon}_i). \tag{A35}$$

When ε_i is longitudinal and ε_f is transverse, we have after some calculations

$$N_1 \sim 16\beta^2(1-\beta)(1-2\beta)\lambda(\mathbf{r}_1 \cdot \boldsymbol{\varepsilon}_f). \tag{A36}$$

The evaluations of $\mathscr{I}_2^V(\mathbf{r}_1, \mathbf{q}_\perp)$ can be carried out similarly. We have

$$\mathscr{I}_2^V(\mathbf{r}_1, \mathbf{q}_\perp) = \frac{1}{4}g^4\omega^{-1}\int_{-\infty}^{\infty} \frac{dq_-}{(2\pi)}\frac{d^4p}{(2\pi)^4}$$

$$\times \frac{\text{Tr}[\not{\varepsilon}_i(\not{p}+\not{r}_1+m)(\gamma_0+\gamma_3)(\not{p}+\not{q}+m)\not{\varepsilon}_f(\not{r}_2+\not{r}_1+\not{p}+\not{q}+m)(\gamma_0+\gamma_3)(\not{r}_2+\not{p}+m)]}{[(r_2+p)^2-m^2][(p+r_1)^2-m^2][(r_2+r_1+p+q)^2-m^2][(p+q)^2-m^2]}. \tag{A37}$$

At this point we note that, because $(\gamma_0 + \gamma_3)^2 = 0$, we can ignore the minus components of all the momenta in the trace. Let us again put $p_+ = -2\omega(1-\beta)$; then

$$\int_{-\infty}^{\infty} \frac{dq_-}{2\pi} \frac{1}{[(r_2+r_1+p+q)^2-m^2+i\varepsilon][(p+q)^2-m^2+i\varepsilon]}$$

$$= \frac{i(2\omega)^{-1}}{[\mathbf{p}_\perp+\mathbf{q}_\perp+(1-\beta)\mathbf{r}_1]^2+m^2-\beta(1-\beta)\lambda^2} \tag{A38}$$

if $0 < \beta < 1$, and is equal to zero otherwise. We therefore restrict β to the interval $(0, 1)$. We also have that, if β is inside this interval,

$$\int_{-\infty}^{\infty} \frac{dp_-}{2\pi} \frac{1}{[(r_2+p)^2-m^2+i\varepsilon][(p+r_1)^2-m^2+i\varepsilon]}$$

$$= \frac{i(2\omega)^{-1}}{(\mathbf{p}_\perp+\beta\mathbf{r}_1)^2+m^2-\beta(1-\beta)\lambda^2}. \tag{A39}$$

From equations (A37) through (A39) we get

$$\mathscr{I}_2^V(\mathbf{r}_1, \mathbf{q}_\perp) = -\frac{g^4}{8\pi}\int \frac{d^2p_\perp}{(2\pi)^2}\int_0^1 d\beta$$

$$\times \frac{N_2}{\{[\mathbf{p}_\perp+\mathbf{q}_\perp+(1-\beta)\mathbf{r}_1]^2+m^2-\beta(1-\beta)\lambda^2\}[(\mathbf{p}_\perp+\beta\mathbf{r}_1)^2+m^2-\beta(1-\beta)\lambda^2]}, \tag{A40}$$

where

$$N_2 = s^{-1} \mathrm{Tr}[\rlap{/}{\epsilon}_i(\rlap{/}{p} + \rlap{/}{r}_1 + m)(\gamma_0 + \gamma_3)(\rlap{/}{p} + \rlap{/}{q} + m)\rlap{/}{\epsilon}_f(\rlap{/}{r}_2 + \rlap{/}{r}_1 + \rlap{/}{p} + \rlap{/}{q} + m)$$

$$\times (\gamma_0 + \gamma_3)(\rlap{/}{r}_2 + \rlap{/}{p} + m)]. \quad \text{(A41)}$$

It remains to evaluate N_2. We have

$$N_2 = -2\beta(1 - \beta)\mathrm{Tr}[\rlap{/}{\epsilon}_i(\rlap{/}{p} + \rlap{/}{r}_1 + m)\rlap{/}{\epsilon}_f(\rlap{/}{r}_2 + \rlap{/}{r}_1 + \rlap{/}{p} + \rlap{/}{q} + m)]$$

$$+ 2(1 - \beta)^2 \mathrm{Tr}[\rlap{/}{\epsilon}_i\rlap{/}{\epsilon}_f(\rlap{/}{r}_2 + \rlap{/}{r}_1 + \rlap{/}{p} + \rlap{/}{q} + m)(\rlap{/}{r}_2 + \rlap{/}{p} + m)]$$

$$+ 2\beta^2 \mathrm{Tr}[\rlap{/}{\epsilon}_i(\rlap{/}{p} + \rlap{/}{r}_1 + m)(\rlap{/}{p} + \rlap{/}{q} - m)\rlap{/}{\epsilon}_f]$$

$$- 2\beta(1 - \beta)\mathrm{Tr}[\rlap{/}{\epsilon}_i(\rlap{/}{p} + \rlap{/}{q} - m)\rlap{/}{\epsilon}_f(\rlap{/}{r}_2 + \rlap{/}{p} - m)]. \quad \text{(A42)}$$

The rest of the calculation is straightforward but tedious. We can simplify the calculations somewhat if we remember that the minus components of all momenta in the trace can be dropped. We get, when ε_i and ε_f are both transverse,

$$N_2 = -8\beta(1 - \beta)\{(p + r_1)_i(r_2 + r_1 + p + q)_j + (r_2 + r_1 + p + q)_i(p + r_1)_j$$

$$- \delta_{ij}[(\mathbf{p}_\perp + \mathbf{r}_1)\cdot(\mathbf{p}_\perp + \mathbf{q}_\perp + \mathbf{r}_1) + m^2]\}$$

$$+ 8(1 - \beta)^2 \{(r_2 + p)_i(r_2 + r_1 + p + q)_j - (r_2 + r_1 + p + q)_i(r_2 + p)_j$$

$$+ \delta_{ij}[\mathbf{p}_\perp\cdot(\mathbf{p}_\perp + \mathbf{q}_\perp + \mathbf{r}_1) + m^2]\}$$

$$+ 8\beta^2\{(p + r_1)_i(p + q)_j - (p + q)_i(p + r_1)_j$$

$$+ \delta_{ij}[(\mathbf{p}_\perp + \mathbf{r}_1)\cdot(\mathbf{p}_\perp + \mathbf{q}_\perp) + m^2]\}$$

$$- 8\beta(1 - \beta)\{(p + q)_i(r_2 + p)_j + (r_2 + p)_i(p + q)_j$$

$$- \delta_{ij}[\mathbf{p}_\perp\cdot(\mathbf{p}_\perp + \mathbf{q}_\perp) + m^2]\}$$

$$= 8\{[(1 - \beta)(r_2 + p) - \beta(p + r_1)]_i$$

$$\times [(1 - \beta)(r_2 + r_1 + p + q) - \beta(p + q)]_j$$

$$- [(1 - \beta)(r_2 + r_1 + p + q) + \beta(p + q)]_i$$

$$\times [(1 - \beta)(r_2 + p) + \beta(p + r_1)]_j$$

$$+ [(1 - \beta)(\mathbf{r}_1 + \mathbf{p}_\perp + \mathbf{q}_\perp) + \beta(\mathbf{p}_\perp + \mathbf{q}_\perp)]$$

$$\times [(1 - \beta)\mathbf{p}_\perp + \beta(\mathbf{p}_\perp + \mathbf{r}_1)]\delta_{ij} + m^2\delta_{ij}\}. \quad \text{(A43)}$$

We simplify equation (A43) by making use of relation (A32) and equations (A12) and obtain, for ε_i and ε_f both transverse,

$$N_2 = 8\{(1 - 2\beta)^2(\beta r_1 + p_\perp)_i[(1 - \beta)r_1 + p_\perp + q_\perp]_j$$

$$- [(1 - \beta)r_1 + p_\perp + q_\perp]_i(\beta r_1 + p_\perp)_j$$

$$+ [p_\perp + q_\perp + (1 - \beta)r_1] \cdot (p_\perp + \beta r_1)\delta_{ij} + m^2 \delta_{ij}\}. \quad \text{(A44)}$$

When ε_i is transverse and ε_f is longitudinal, we have after some calculations

$$N_2 = 16\beta(1 - \beta)(1 - 2\beta)\lambda(\mathbf{Q} \cdot \varepsilon_i), \quad \text{(A45)}$$

where

$$\mathbf{Q} = \tfrac{1}{2}q_\perp + (\tfrac{1}{2} - \beta)r_1. \quad \text{(A46)}$$

When ε_i is longitudinal and ε_f is transverse, we have after some calculations

$$N_2 = -16\beta(1 - \beta)(1 - 2\beta)\lambda(\mathbf{Q} \cdot \varepsilon_f). \quad \text{(A47)}$$

By equations (9.36), (A28), and (A40), we have

$$\mathscr{I}^V(\mathbf{r}_1, \mathbf{q}_\perp) = \frac{g^4}{\pi} \int \frac{d^2 p_\perp}{(2\pi)^2} \int_0^1 \frac{d\beta}{(p_\perp + \beta r_1)^2 + m^2 - \beta(1 - \beta)\lambda^2}$$

$$\times \left\{ \frac{\tfrac{1}{8}N_1}{(p_\perp - \beta r_1)^2 + m^2 - \beta(1 - \beta)\lambda^2} \right.$$

$$\left. - \frac{\tfrac{1}{8}N_2}{[p_\perp + q_\perp + (1 - \beta)r_1]^2 + m^2 - \beta(1 - \beta)\lambda^2} \right\}$$

$$= \frac{g^4}{\pi} \int \frac{d^2 p_\perp}{(2\pi)^2} \int_0^1 \frac{d\beta}{(p_\perp + \beta r_1)^2 + m^2 - \beta(1 - \beta)\lambda^2}$$

$$\times \left\{ \frac{\tfrac{1}{8}N_1 - [(p_\perp - \beta r_1)^2 + m^2 - \beta(1 - \beta)\lambda^2]\delta_{ij}}{(p_\perp - \beta r_1)^2 + m^2 - \beta(1 - \beta)\lambda^2} \right.$$

$$\left. - \frac{\tfrac{1}{8}N_2 - [p_\perp + q_\perp + (1 - \beta)r_1]\delta_{ij} - [m^2 - \beta(1 - \beta)\lambda^2]\delta_{ij}}{[p_\perp + q_\perp + (1 - \beta)r_1]^2 + m^2 - \beta(1 - \beta)\lambda^2} \right\}.$$

$$\text{(A48)}$$

Making the change of variable $p_\perp \to p_\perp - \tfrac{1}{2}(q_\perp + r_1)$ for the second term in the braces in equation (A48), we get, if both ε_i and ε_f are transverse,

$$\mathscr{I}^V(\mathbf{r}_1, \mathbf{q}_\perp) = \frac{g^4}{4\pi^3} \int d^2 p_\perp \int_0^1 d\beta$$

$$\times \left\{ \frac{4\beta(1-\beta)(\beta^2 \mathbf{r}_{1i}\mathbf{r}_{1j} - \mathbf{p}_{\perp i}\mathbf{p}_{\perp j}) - [2\beta^2 \mathbf{r}_1^2 - \beta(1-\beta)\lambda^2]\delta_{ij}}{[(\mathbf{p}_\perp + \beta \mathbf{r}_1)^2 + m^2 - \beta(1-\beta)\lambda^2][(\mathbf{p}_\perp - \beta \mathbf{r}_1)^2 + m^2 - \beta(1-\beta)\lambda^2]} \right.$$

$$\left. - \text{preceding term with } \beta \mathbf{r}_1 \to \mathbf{Q} \right\}. \quad (A49)$$

Equation (A49) can be written in another way. By introducing a Feynman parameter x and carrying out the integration over $\mathbf{p}_\perp$, we obtain for both ε_i and ε_f transverse,

$$\mathscr{I}^V(\mathbf{r}_1, \mathbf{q}_\perp) = \frac{g^4}{2\pi^2} \int_0^1 dx \int_0^1 d\beta$$

$$\times \left\{ \frac{8\beta^3(1-\beta)x(1-x)\mathbf{r}_{1i}\mathbf{r}_{1j} - \beta^2 \mathbf{r}_1^2[1 - 8\beta(1-\beta)(x - \frac{1}{2})^2]\delta_{ij} + \beta(1-\beta)\lambda^2 \delta_{ij}}{4\beta^2 x(1-x)\mathbf{r}_1^2 + m^2 - \beta(1-\beta)\lambda^2} \right.$$

$$\left. - \text{preceding term with } \beta \mathbf{r}_1 \to \mathbf{Q} \right\}. \quad (A50)$$

For other polarizations we replace the numerator in equation (A50) by the expressions in relations (A34), (A35), and (A36) divided by 4.

Reference

[1] H. Cheng and T. T. Wu, *Physical Review* 182 (1969): 1899, *Physical Review D* 1 (1970): 459.

Appendix B

In this appendix we consider in some detail, for the tower diagrams studied in chapters 6 and 11, the special case of forward scattering, where $\mathbf{r}_1 = 0$. We also discuss generalizations to nonforward directions. In particular, we give a derivation of expression (6.23).

B.1 Kernel in the Forward Direction

Let

$$K_0(\mathbf{q}_\perp, \mathbf{q}'_\perp) = [4g^4(2\pi)^{-3}]^{-1} K(0, \mathbf{q}_\perp, \mathbf{q}'_\perp), \tag{B1}$$

where K is defined by equation (11.37). It follows from equation (11.42) that the two expressions for K_0 are

$$K_0(\mathbf{q}_\perp, \mathbf{q}'_\perp) = \tfrac{1}{4}(2\pi)^{-1} \int d^2 p_\perp \int_0^1 dx [x\mathbf{p}_\perp^2 + (1-x)(\mathbf{p}_\perp - \mathbf{q}_\perp)^2 + m^2]^{-1}$$

$$\times \left\{ -\frac{[(\mathbf{p}_\perp - \mathbf{q}_\perp)^2 + m^2]\,\mathrm{Tr}[(\not{p}_\perp - m)(\not{p}_\perp + m)]}{x\mathbf{p}_\perp^2 + (1-x)(\mathbf{p}_\perp - \mathbf{q}_\perp)^2 + m^2} \right.$$

$$\left. - \frac{\mathrm{Tr}[(\not{p}_\perp - m)(\not{p}_\perp - \not{q}'_\perp + m)(\not{p}_\perp - \not{q}_\perp - \not{q}'_\perp - m)(\not{p}_\perp - \not{q}_\perp + m)]}{x(\mathbf{p}_\perp - \mathbf{q}'_\perp)^2 + (1-x)(\mathbf{p}_\perp - \mathbf{q}_\perp - \mathbf{q}'_\perp)^2 + m^2} \right\} \tag{B2}$$

and

$$K_0(\mathbf{q}_\perp, \mathbf{q}'_\perp) = (2\pi)^{-1} \int d^2 p_\perp \left\{ 1 - \frac{1}{4} \right.$$

$$\times \frac{\mathrm{Tr}[(\not{p}_\perp - m)(\not{p}_\perp - \not{q}'_\perp + m)(\not{p}_\perp - \not{q}_\perp - \not{q}'_\perp - m)(\not{p}_\perp - \not{q}_\perp + m)]}{(\mathbf{p}_\perp^2 + m^2)[(\mathbf{p}_\perp - \mathbf{q}_\perp - \mathbf{q}'_\perp)^2 + m^2] - [(\mathbf{p}_\perp - \mathbf{q}_\perp)^2 + m^2][(\mathbf{p}_\perp - \mathbf{q}'_\perp)^2 + m^2]}$$

$$\left. \times \ln \frac{(\mathbf{p}_\perp^2 + m^2)[(\mathbf{p}_\perp - \mathbf{q}_\perp - \mathbf{q}'_\perp)^2 + m^2]}{[(\mathbf{p}_\perp - \mathbf{q}_\perp)^2 + m^2][(\mathbf{p}_\perp - \mathbf{q}'_\perp)^2 + m^2]} \right\}. \tag{B3}$$

These two expressions can be usefully combined in the form

$$K_0(\mathbf{q}_\perp, \mathbf{q}'_\perp) = (2\pi)^{-1} \int d^2 p_\perp \int_0^1 dx \left\{ 1 - \frac{1}{4} \right.$$

$$\left. \times \frac{\mathrm{Tr}[(\not{p}_\perp - m)(\not{p}_\perp - \not{q}'_\perp + m)(\not{p}_\perp - \not{q}_\perp - \not{q}'_\perp - m)(\not{p}_\perp - \not{q}_\perp + m)]}{[x\mathbf{p}_\perp^2 + (1-x)(\mathbf{p}_\perp - \mathbf{q}_\perp)^2 + m^2][x(\mathbf{p}_\perp - \mathbf{q}'_\perp)^2 + (1-x)(\mathbf{p}_\perp - \mathbf{q}_\perp - \mathbf{q}'_\perp)^2 + m^2]} \right\}. \tag{B4}$$

If the trace of the γ-matrices is written explicitly, the result is

$$K_0(\mathbf{q}_\perp, \mathbf{q}'_\perp) = (2\pi)^{-1} \int d^2 p_\perp \int_0^1 dx$$

$$\times \left\{ 1 - \frac{(\mathbf{p}_\perp^2 + m^2)^2 - (\mathbf{p}_\perp^2 + m^2)[2\mathbf{p}_\perp \cdot (\mathbf{q}_\perp + \mathbf{q}'_\perp) - (\mathbf{q}_\perp^2 + \mathbf{q}_\perp \cdot \mathbf{q}'_\perp + \mathbf{q}'^2_\perp)] + 2(\mathbf{p}_\perp \cdot \mathbf{q}_\perp)(\mathbf{p}_\perp \cdot \mathbf{q}'_\perp) - (\mathbf{p}_\perp \cdot \mathbf{q}_\perp)\mathbf{q}'^2_\perp - (\mathbf{p}_\perp \cdot \mathbf{q}'_\perp)\mathbf{q}_\perp^2}{[x\mathbf{p}_\perp^2 + (1-x)(\mathbf{p}_\perp - \mathbf{q}_\perp)^2 + m^2][x(\mathbf{p}_\perp - \mathbf{q}'_\perp)^2 + (1-x)(\mathbf{p}_\perp - \mathbf{q}_\perp - \mathbf{q}'_\perp)^2 + m^2]} \right\}.$$

$$\text{(B5)}$$

This is the starting point of the present investigation.

Equation (B5) fails to exhibit the symmetry property (11.43). To restore this symmetry explicitly, we introduce a second Feynman parameter y to combine the two denominators:

$$K_0(\mathbf{q}_\perp, \mathbf{q}'_\perp) = (2\pi)^{-1} \int d^2 p_\perp \int_0^1 dx \int_0^1 dy (1 - N_0/D_0^2), \tag{B6}$$

where N_0 is the numerator that appears in equation (B5) and D_0 is given by

$$D_0 = xy\mathbf{p}_\perp^2 + (1-x)y(\mathbf{p}_\perp - \mathbf{q}_\perp)^2 + x(1-y)(\mathbf{p}_\perp - \mathbf{q}'_\perp)^2$$

$$\qquad + (1-x)(1-y)(\mathbf{p}_\perp - \mathbf{q}_\perp - \mathbf{q}'_\perp)^2 + m^2$$

$$\quad = \mathbf{p}_\perp^2 - 2(1-x)\mathbf{p}_\perp \cdot \mathbf{q}_\perp - 2(1-y)\mathbf{p}_\perp \cdot \mathbf{q}'_\perp + (1-x)y\mathbf{q}_\perp^2$$

$$\qquad + x(1-y)\mathbf{q}'^2_\perp + (1-x)(1-y)(\mathbf{q}_\perp + \mathbf{q}'_\perp)^2 + m^2. \tag{B7}$$

We reverse the order of integration and rewrite equation (B6) in the form

$$K_0(\mathbf{q}_\perp, \mathbf{q}'_\perp) = (2\pi)^{-1} \int_0^1 dx \int_0^1 dy \int d^2 p_\perp (1 - N_0/D_0^2). \tag{B8}$$

Because of equation (B7), we write

$$\mathbf{p}_\perp = \delta \mathbf{p}_\perp + \mathbf{p}'_\perp, \tag{B9}$$

where

$$\delta \mathbf{p}_\perp = (1-x)\mathbf{q}_\perp + (1-y)\mathbf{q}'_\perp. \tag{B10}$$

In changing to this new variable $\mathbf{p}'_\perp$, we must take into account the fact that

$$\int d^2 p_\perp (1 - N_0/D_0^2)$$

is linearly divergent. More precisely, this shift gives the contribution

$$-\int_0^1 dx \int_0^1 dy\, \delta\mathbf{p}_\perp \cdot [(\mathbf{q}_\perp + \mathbf{q}'_\perp) - 2\delta\mathbf{p}_\perp]$$

$$= \int_0^1 dx \int_0^1 dy[(1-x)(1-2x)\mathbf{q}_\perp^2 + (1-y)(1-2y)\mathbf{q}_\perp'^2]$$

$$= \tfrac{1}{6}(\mathbf{q}_\perp^2 + \mathbf{q}_\perp'^2). \tag{B11}$$

With this contribution properly included and after a tedious calculation, symmetric integration over $p'_\perp$ yields

$$K_0(\mathbf{q}_\perp, \mathbf{q}'_\perp) = \tfrac{1}{6}(\mathbf{q}_\perp^2 + \mathbf{q}_\perp'^2) - (2\pi)^{-1} \int_0^1 dx \int_0^1 dy \int d^2 p'_\perp$$

$$\times [\mathbf{p}_\perp'^2 + x(1-x)\mathbf{q}_\perp^2 + y(1-y)\mathbf{q}_\perp'^2 + m^2]^{-2}$$

$$\times \{(\mathbf{p}_\perp'^2 + m^2)[(1 - 6x + 6x^2)\mathbf{q}_\perp^2 + 2(1-2x)(1-2y)\mathbf{q}_\perp \cdot \mathbf{q}'_\perp$$

$$+ (1 - 6y + 6y^2)\mathbf{q}_\perp'^2]$$

$$+ m^2[2x(1-x)\mathbf{q}_\perp^2 - (1-2x)(1-2y)\mathbf{q}_\perp \cdot \mathbf{q}'_\perp + 2y(1-y)\mathbf{q}_\perp'^2]$$

$$- x(1-x)(1-2x)(1-2y)\mathbf{q}_\perp^2(\mathbf{q}_\perp \cdot \mathbf{q}'_\perp) - [x(1-x) + y(1-y)]\mathbf{q}_\perp^2\mathbf{q}_\perp'^2$$

$$+ 4x(1-x)y(1-y)(\mathbf{q}_\perp \cdot \mathbf{q}'_\perp)^2 - y(1-y)(1-2x)(1-2y)\mathbf{q}_\perp'^2(\mathbf{q}_\perp \cdot \mathbf{q}'_\perp)\}. \tag{B12}$$

At this stage, simplification can be achieved by noticing that the denominator is not changed by the replacement $x \to 1 - x$. Thus a number of terms in equation (B12) do not contribute, and

$$K_0(\mathbf{q}_\perp, \mathbf{q}'_\perp) = \tfrac{1}{6}(\mathbf{q}_\perp^2 + \mathbf{q}_\perp'^2) - (2\pi)^{-1} \int_0^1 dx \int_0^1 dy \int d^2 p'_\perp$$

$$\times [\mathbf{p}_\perp'^2 + x(1-x)\mathbf{q}_\perp^2 + y(1-y)\mathbf{q}_\perp'^2 + m^2]^{-2}$$

$$\times \{(\mathbf{p}_\perp'^2 + m^2)[(1 - 6x + 6x^2)\mathbf{q}_\perp^2 + (1 - 6y + 6y^2)\mathbf{q}_\perp'^2]$$

$$+ 2m^2[x(1-x)\mathbf{q}_\perp^2 + y(1-y)\mathbf{q}_\perp'^2] - [x(1-x) + y(1-y)]\mathbf{q}_\perp^2\mathbf{q}_\perp'^2$$

$$+ 4x(1-x)y(1-y)(\mathbf{q}_\perp \cdot \mathbf{q}'_\perp)^2\}. \tag{B13}$$

The $p'_\perp$ integral on the right-hand side of equation (B13) is still logarithmically divergent. This divergence does not cause any trouble because

$$\int_0^1 dx(1 - 6x + 6x^2) = 0. \tag{B14}$$

It is now a straightforward task to carry out the integration over $p'_\perp$. We get

$$K_0(\mathbf{q}_\perp, \mathbf{q}'_\perp)$$

$$= \tfrac{1}{6}(\mathbf{q}_\perp^2 + \mathbf{q}_\perp'^2) + \tfrac{1}{2}\int_0^1 dx \int_0^1 dy([(1 - 6x + 6x^2)\mathbf{q}_\perp^2 + (1 - 6y + 6y^2)\mathbf{q}_\perp'^2]$$

$$\times \ln[x(1 - x)\mathbf{q}_\perp^2 + y(1 - y)\mathbf{q}_\perp'^2 + m^2]$$

$$+ [x(1 - x)\mathbf{q}_\perp^2 + y(1 - y)\mathbf{q}_\perp'^2 + m^2]^{-1}$$

$$\times \{[x(1 - x)\mathbf{q}_\perp^2 + y(1 - y)\mathbf{q}_\perp'^2]$$

$$\times [(1 - 6x + 6x^2)\mathbf{q}_\perp^2 + (1 - 6y + 6y^2)\mathbf{q}_\perp'^2 - 2m^2]$$

$$+ [x(1 - x) + y(1 - y)]\mathbf{q}_\perp^2\mathbf{q}_\perp'^2 - 4x(1 - x)y(1 - y)(\mathbf{q}_\perp \cdot \mathbf{q}'_\perp)^2\})$$

$$= \tfrac{1}{6}(\mathbf{q}_\perp^2 + \mathbf{q}_\perp'^2) - \int_0^1 dx \int_0^1 dy[x(1 - x)\mathbf{q}_\perp^2 + y(1 - y)\mathbf{q}_\perp'^2]$$

$$+ \tfrac{1}{2}\int_0^1 dx \int_0^1 dy[x(1 - x)\mathbf{q}_\perp^2 + y(1 - y)\mathbf{q}_\perp'^2 + m^2]^{-1}$$

$$\times \{-x(1 - x)(1 - 2x)^2(\mathbf{q}_\perp^2)^2 - y(1 - y)(1 - 2y)^2(\mathbf{q}_\perp'^2)^2$$

$$+ [x(1 - x)\mathbf{q}_\perp^2 + y(1 - y)\mathbf{q}_\perp'^2][(1 - 2x)^2\mathbf{q}_\perp^2 + (1 - 2y)^2\mathbf{q}_\perp'^2]$$

$$+ [x(1 - x) + y(1 - y)]\mathbf{q}_\perp^2\mathbf{q}_\perp'^2 - 4x(1 - x)y(1 - y)(\mathbf{q}_\perp \cdot \mathbf{q}'_\perp)^2\}. \quad \text{(B15)}$$

Accordingly, we get the desired answer:

$$K_0(\mathbf{q}_\perp, \mathbf{q}'_\perp) = \int_0^1 dx \int_0^1 dy$$

$$\times \frac{[x(1 - x) + y(1 - y)]\mathbf{q}_\perp^2\mathbf{q}_\perp'^2 - 2x(1 - x)y(1 - y)[2\mathbf{q}_\perp^2\mathbf{q}_\perp'^2 + (\mathbf{q}_\perp \cdot \mathbf{q}'_\perp)^2]}{x(1 - x)\mathbf{q}_\perp^2 + y(1 - y)\mathbf{q}_\perp'^2 + m^2}.$$

$$\text{(B16)}$$

This form exhibits explicitly the properties

$$K_0(\mathbf{q}_\perp, \mathbf{q}'_\perp) = K_0(\mathbf{q}'_\perp, \mathbf{q}_\perp) \tag{B17}$$

and

$$K_0(\mathbf{q}_\perp, 0) = K_0(0, \mathbf{q}'_\perp) = 0, \tag{B18}$$

which are the special cases of equations (11.43) and (11.44) for $\mathbf{r}_1 = 0$.

The integrations in the right-hand side of equation (B16) can be carried out explicitly in terms of Clausen's integral. Because the resulting formula is not needed, however, we refer the reader to the literature [1].

B.2 Spectrum of the Kernel

In this section we study the spectrum of the kernel at $\mathbf{r}_1 = 0$. Let θ be the angle between $\mathbf{q}_\perp$ and $\mathbf{q}'_\perp$. Then the K_0 of equation (B16) can be written

$$K_0(\mathbf{q}_\perp, \mathbf{q}'_\perp) = \mathbf{q}_\perp^2 \mathbf{q}'^2_\perp \int_0^1 dx \int_0^1 dy$$

$$\times \frac{[x(1-x) + y(1-y)] - 2x(1-x)y(1-y)(2 + \cos^2\theta)}{x(1-x)\mathbf{q}_\perp^2 + y(1-y)\mathbf{q}'^2_\perp + m^2}. \quad (B19)$$

We are interested only in applying this kernel to functions that are rotationally invariant. We can therefore average over θ in equation (B19) to get

$$\langle K_0(\mathbf{q}_\perp, \mathbf{q}'_\perp)\rangle_\theta$$

$$= zz' \int_0^1 dx \int_0^1 dy \frac{x(1-x) + y(1-y) - 5x(1-x)y(1-y)}{x(1-x)z + y(1-y)z' + m^2}, \quad (B20)$$

where

$$z = \mathbf{q}_\perp^2 \quad \text{and} \quad z' = \mathbf{q}'^2_\perp. \quad (B21)$$

Define the kernel by (see identity (6.16))

$$\mathcal{K}_0(z, z') = (z + \lambda^2)^{-1}(z' + \lambda^2)^{-1}\langle K_0(\mathbf{q}_\perp, \mathbf{q}'_\perp)\rangle_\theta$$

$$= \frac{z}{z + \lambda^2} \frac{z'}{z' + \lambda^2} \int_0^1 dx \int_0^1 dy$$

$$\times \frac{x(1-x) + y(1-y) - 5x(1-x)y(1-y)}{x(1-x)z + y(1-y)z' + m^2} \quad (B22)$$

and the corresponding operator $\mathcal{K}_0$ by

$$(\mathcal{K}_0 f)(z) = \int_0^\infty dz' \mathcal{K}_0(z, z')f(z'). \quad (B23)$$

For the sake of mathematical rigor, we let $f(z)$ be elements of the L_2-space; that is, we consider those $f(z)$ that satisfy

$$\int_0^\infty |f(z)|^2 \, dz < \infty. \tag{B24}$$

We see later that $\mathcal{K}_0$ is a bounded operator on L_2. Therefore, as a consequence of

$$\mathcal{K}_0(z, z') = \mathcal{K}_0(z', z), \tag{B25}$$

the spectrum $\mathcal{S}$ of $\mathcal{K}_0$ is a real, bounded, closed set. Let μ_0 be the lowest upper bound of this set. It is the purpose of this section to calculate μ_0. Our result here is simply

$$\mu_0 = 11\pi^3/64, \tag{B26}$$

independent of m and λ.

The remainder of this section is devoted to the derivation of equation (B26). We first study the solvable special case $m = \lambda = 0$, where

$$\mathcal{K}_{00}(z, z') = \mathcal{K}_0(z, z')|_{m=\lambda=0}$$

$$= \int_0^1 dx \int_0^1 dy \frac{x(1-x) + y(1-y) - 5x(1-x)y(1-y)}{x(1-x)z + y(1-y)z'}. \tag{B27}$$

Let

$$z = e^\xi, \qquad z' = e^{\xi'}, \tag{B28}$$

and

$$f(z) = e^{-\xi/2} g(\xi). \tag{B29}$$

The reason for using equation (B29) is that relation (B24) is equivalent to

$$\int_{-\infty}^\infty |g(\xi)|^2 \, d\xi < \infty. \tag{B30}$$

In the ξ-space, we need to study the kernel

$$\exp[(\xi + \xi')/2] \int_0^1 dx \int_0^1 dy \frac{x(1-x) + y(1-y) - 5x(1-x)y(1-y)}{x(1-x)e^\xi + y(1-y)e^{\xi'}}$$

$$= \int_0^1 dx \int_0^1 dy \frac{x(1-x) + y(1-y) - 5x(1-x)y(1-y)}{x(1-x)\exp[(\xi - \xi')/2] + y(1-y)\exp[-(\xi - \xi')/2]}, \tag{B31}$$

which is a function of $\xi - \xi'$ only. The Fourier transform of the right-hand side of equation (B31) is

$$\int_{-\infty}^{\infty} d\xi \exp(i\xi\tau) \int_0^1 dx \int_0^1 dy \frac{x(1-x) + y(1-y) - 5x(1-x)y(1-y)}{x(1-x)e^{\xi/2} + y(1-y)e^{-\xi/2}}$$

$$= \pi \operatorname{sech} \pi\tau \int_0^1 dx \int_0^1 dy[x(1-x) + y(1-y) - 5x(1-x)y(1-y)]$$

$$\times [x(1-x)y(1-y)]^{-1/2} \exp\{-i\tau[\ln x(1-x) - \ln y(1-y)]\}$$

$$= \pi \operatorname{sech} \pi\tau \left\{ 2 \operatorname{Re} \frac{[\Gamma(\frac{3}{2} - i\tau)]^2 [\Gamma(\frac{1}{2} + i\tau)]^2}{\Gamma(3 - 2i\tau)\Gamma(1 + 2i\tau)} - 5\frac{[\Gamma(\frac{3}{2} - i\tau)]^2 [\Gamma(\frac{3}{2} + i\tau)]^2}{\Gamma(3 - 2i\tau)\Gamma(3 + 2i\tau)} \right\}$$

$$= \frac{(\pi \operatorname{sech} \pi\tau)^3}{2\pi\tau \operatorname{csch} 2\pi\tau} \left(\frac{1}{2} \operatorname{Re} \frac{1 - 2i\tau}{2 - 2i\tau} - \frac{5}{16} \frac{1 + 4\tau^2}{4 + 4\tau^2} \right)$$

$$= \frac{\pi^2}{64} \frac{11 + 12\tau^2}{1 + \tau^2} \frac{\sinh \pi\tau}{\tau \cosh^2 \pi\tau}. \tag{B32}$$

For real values of τ, the right-hand side of equation (B32) takes on all real values between 0 and $11\pi^3/64$. Because this is bounded, it follows from relation (B30) that, when $m = \lambda = 0$, the spectrum of $\mathcal{K}_{00}$ is the closure of the values taken by the right-hand side of equation (B32); that is, the spectrum of $\mathcal{K}_{00}$ is

$$[0, \ 11\pi^3/64]. \tag{B33}$$

Next we study the case $m \neq 0$ and $\lambda \neq 0$. We first note that

$$x(1-x) + y(1-y) - 5x(1-x)y(1-y) > 0 \tag{B34}$$

for $0 < x < 1$ and $0 < y < 1$. Accordingly, it follows from expression (B33) that

$$\int_0^{\infty} |(\mathcal{K}_0 f)(z)|^2 \, dz \leqslant (11\pi^3/64)^2 \int_0^{\infty} |f(z)|^2 \, dz, \tag{B35}$$

and hence

$$\mu_0 \leqslant 11\pi^3/64 \tag{B36}$$

for all m and λ. Relation (B35) further implies that $\mathcal{K}_0$ is a bounded operator on L_2.

On the other hand, because

$$\mu_0 = \sup \frac{\int_0^\infty dz \int_0^\infty dz'\, f(z)f(z')\mathcal{K}_0(z,z')}{\int_0^\infty [f(z)]^2\, dz} \tag{B37}$$

over all real nonzero $f(z)$ that satisfy relation (B24), we can obtain a lower bound for μ_0 by trying some $f(z)$. In particular, we choose, for all $\Lambda > 1$,

$$f(z,\Lambda) = \begin{cases} z^{-3/2}(z + \lambda^2) & \text{for } 1 \leqslant z \leqslant \Lambda, \\ 0 & \text{otherwise.} \end{cases} \tag{B38}$$

Then

$$\mu_0 \geqslant \sup_\Lambda \left[\ln \Lambda + 2\lambda^2(1 - \Lambda^{-1}) + \tfrac{1}{2}\lambda^4(1 - \Lambda^{-2})\right]^{-1}$$

$$\times \int_0^\infty dz \int_0^\infty dz'\, f(z,\Lambda)f(z',\Lambda)\mathcal{K}_0(z,z'). \tag{B39}$$

The reason for this choice of $f(z,\Lambda)$ is as follows. Because the right-hand side of equation (B32) has a maximum at $\tau = 0$, the eigenfunction that corresponds to μ_0 is, by equations (B28) and (B29), simply $z^{-1/2}$, which of course does not satisfy relation (B24) and holds only for $m = \lambda = 0$. The $f(z,\Lambda)$ of equation (B38) is essentially the product of this function $z^{-1/2}$ with the inverse of the factor $z/(z + \lambda^2)$ that appears in equation (B22). By explicit calculation, it is easy to verify that

$$\lim_{\Lambda \to \infty} (\ln \Lambda)^{-1} \int_1^\Lambda dz \int_1^\Lambda dz'\, z^{-1/2} z'^{-1/2} \mathcal{K}_0(z,z')\big|_{m=\lambda=0} = 11\pi^3/64. \tag{B40}$$

If equation (B40) is substituted into relation (B39), we get in particular

$$\mu_0 \geqslant 11\pi^3/64 - \lim_{\Lambda \to \infty} (\ln \Lambda)^{-1} I(\Lambda, m), \tag{B41}$$

where, by equation (B22),

$$I(\Lambda, m) = \int_1^\Lambda dz \int_1^\Lambda dz'\, z^{-1/2} z'^{-1/2} \int_0^1 dx \int_0^1 dy [x(1 - x) + y(1 - y)$$

$$- 5x(1 - x)y(1 - y)]\{[x(1 - x)z + y(1 - y)z']^{-1}$$

$$- [x(1 - x)z + y(1 - y)z' + m^2]^{-1}\}. \tag{B42}$$

Note that $I(\Lambda, m)$ does not depend on λ and that the integrand of equation (B42) is nonnegative because of relation (B34).

It can be verified that

$$I(m) = \int_1^\infty dz \int_1^\infty dz'\, z^{-1/2} z'^{-1/2} \int_0^1 dx \int_0^1 dy [x(1-x) + y(1-y)]$$
$$- 5x(1-x)y(1-y)]\{[x(1-x)z + y(1-y)z']^{-1}$$
$$- [x(1-x)z + y(1-y)z' + m^2]^{-1}\} \quad (B43)$$

exists. Therefore

$$\lim_{\Lambda\to\infty} I(\Lambda, m) = I(m),$$

and hence by relation (B41)

$$\mu_0 \geqslant 11\pi^3/64. \tag{B44}$$

The required answer (equation (B26)) then follows from relations (B36) and (B44).

B.3   Eigenfunctions for μ near $\mu_0 = 11\pi^3/64$

Having found the spectrum as given by expression (B33), we study next the continuum eigenfunctions near the upper end, viz., the $\phi(z, \mu)$ of

$$\int_0^\infty \mathscr{K}_0(z, z')\phi(z', \mu)\, dz' = \mu\phi(z, \mu), \tag{B45}$$

with $\mathscr{K}_0(z, z')$ given by equation (B22) and μ close to the μ_0 of equation (B26). The considerations of this section apply when

$$\lambda > 0 \quad \text{and/or} \quad m > 0. \tag{B46}$$

We consider first the case $\mu = \mu_0$. With

$$\phi_0(z) = \phi(z, \mu_0), \tag{B47}$$

the integral equation under consideration is

$$\int_0^\infty \mathscr{K}_0(z, z')\phi_0(z')\, dz' = \mu_0\phi_0(z). \tag{B48}$$

This $\phi_0(z)$ is not normalized. Because μ_0 is the least upper bound of the spectrum, $\phi_0(z)$ can be chosen to be positive:

$$\phi_0(z) \geqslant 0. \tag{B49}$$

A central problem here is the determination of the asymptotic behavior of $\phi_0(z)$ for large z. The importance of this point is as follows. For μ near μ_0, the behavior of $\phi(z, \mu)$ is given by

$$\phi(z, \mu) \sim \text{const } \phi_0(z)$$

for $z = O(1)$ and by the sum of two exponentials for $z = O[(\mu_0 - \mu)^{-1/2}]$ (see relation (B64)). Because the normalization is controlled by these exponentials, the asymptotic behavior of $\phi_0(z)$ for large z determines, through the connection formula, the magnitude of the normalized eigenfunctions for μ near μ_0. It is this magnitude that is used in the next section to ascertain the power of $\ln s$ in relation (6.23).

With the $\mathcal{K}_{00}(z, z')$ of equation (B27), define

$$\Delta_0(z, z') = \mathcal{K}_{00}(z, z') - \mathcal{K}_0(z, z'). \tag{B50}$$

Then, under relation (B46),

$$\Delta_0(z, z') > 0 \tag{B51}$$

for all nonnegative z and z'. With equation (B50), the integral equation (B48) can be rewritten as

$$\int_0^\infty \mathcal{K}_{00}(z, z')\phi_0(z')\, dz' - \mu_0\phi_0(z) = \phi_1(z), \tag{B52}$$

where

$$\phi_1(z) = \int_0^\infty \Delta_0(z, z')\phi_0(z')\, dz'. \tag{B53}$$

By relations (B49) and (B51),

$$\phi_1(z) > 0 \tag{B54}$$

for all nonnegative z.

In order to find the asymptotic behavior of $\phi_0(z)$ for large z, we use the method of Mellin transform:

$$\Phi_0(\zeta) = \int_0^\infty dz\, z^{-1/2-\zeta}\phi_0(z),$$

$$\Phi_1(\zeta) = \int_0^\infty dz\, z^{-1/2-\zeta} \phi_1(z). \tag{B55}$$

With reference to section B.2, this Mellin transform is closely related to the change of variable in equations (B28) followed by a Fourier transform with respect to the new variable ξ. This method of Mellin transform is used extensively in appendix C. The power of this method stems from the fact that $\phi_0(z)$ asymptotically approaches $z^{-1/2}(\ln z)^n$ for $z \to \infty$ if and only if $\Phi_0(\zeta) \sim \text{const}\, \zeta^{-n-1}$ as $\zeta \to 0^+$ and there is no other singularity on the imaginary axis.

By equations (B26), (B27), (B32), and (B55), the Mellin transform of equation (B52) is

$$\frac{\pi^3}{64}\left(\frac{11 - 12\zeta^2}{1 - \zeta^2}\, \frac{\sin \pi\zeta}{\pi\zeta \cos^2 \pi\zeta} - 11\right)\Phi_0(\zeta) = \Phi_1(\zeta). \tag{B56}$$

It follows from equations (B22), (B27), and (B50) that $\Phi_1(\zeta)$ is analytic at $\zeta = 0$. Furthermore, relation (B54) implies that

$$\Phi_1(0) = \int_0^\infty dz\, z^{-1/2} \phi_1(z) > 0. \tag{B57}$$

Expansion of equation (B56) about $\zeta = 0$ then gives

$$\Phi_0(\zeta) \sim C_0 \zeta^{-2}, \tag{B58}$$

where C_0 is a positive constant:

$$C_0 = 64\pi^{-3} A_0 \Phi_1(0), \tag{B59}$$

with

$$A_0 = (\tfrac{55}{6}\pi^2 - 1)^{-1}. \tag{B60}$$

Therefore the asymptotic behavior of $\phi_0(z)$ is given by

$$\phi_0(z) = C_0 z^{-1/2}[\ln z + O(1)] \tag{B61}$$

as $z \to \infty$. This is the desired answer.

So far in this section, only the case $\mu = \mu_0$ has been considered. Now let μ be slightly smaller than μ_0. By equation (B32), define a positive τ such that

$$\frac{\pi^2}{64}\, \frac{11 + 12\tau^2}{1 + \tau^2}\, \frac{\sinh \pi\tau}{\tau \cosh^2 \pi\tau} = \mu. \tag{B62}$$

When μ is close to μ_0, τ is small and

$$\tau \sim [A_0(\mu_0 - \mu)]^{1/2}. \tag{B63}$$

For the special case $m = \lambda = 0$ solved explicitly in section B.2, for each μ there are two continuum eigenfunctions, namely, $z^{-1/2+i\tau}$ and $z^{-1/2-i\tau}$. In contrast, when relations (B46) hold, there is only one continuum eigenfunction for each value of μ. Without loss of generality, this eigenfunction can be chosen to be real and is given approximately by

$$\phi(z, \mu) \sim \begin{cases} C_0^{-1} C_0(\mu) \phi_0(z) & \text{for } z \ll \tau^{-1}, \\ C_1(\mu) \operatorname{Im} z^{-1/2+i\tau} e^{i\theta(\mu)} & \text{for } z \gg 1, \end{cases} \tag{B64}$$

where $C_0(\mu)$ is chosen to be positive, whereas $C_1(\mu)$ and $\theta(\mu)$ are real. In relation (B64), the 1 means $\max(m^2, \lambda^2)$. By equation (B61), matching in the overlapping region gives

$$C_0(\mu) [\ln z + O(1)] = C_1(\mu) \sin[\tau \ln z + \theta(\mu)]. \tag{B65}$$

Because τ is small, this is satisfied by

$$\theta(\mu) = O(\tau) \tag{B66}$$

and

$$C_0(\mu) \sim \tau C_1(\mu). \tag{B67}$$

This factor of τ plays an important role in the next section.

B.4 High-Energy Behavior of Tower Diagrams in the Forward Direction

Let the coefficients $C_0(\mu)$ and $C_1(\mu)$ of relation (B64) be chosen so that the eigenfunctions $\phi(z, \mu)$ are normalized in the sense that

$$\int_0^\infty dz \, \phi(z, \mu) \phi(z, \mu') = \delta(\mu - \mu'). \tag{B68}$$

(If this normalization were used in section B.3, then the only equation that would need modification would be equation (B47).) With equation (B68), the spectral decomposition of the kernel $\mathcal{K}_0(z, z')$ is

$$\mathcal{K}_0(z, z') = \int_{\mathscr{S}} \mu \, d\mu \, \phi(z, \mu) \phi(z', \mu), \tag{B69}$$

where $\mathscr{S}$ is the spectrum studied in section B.2. Using equation (B69), we proceed to determine the high-energy behavior of the tower diagrams in the forward direction from identity (6.22), where J is given by identity (6.19):

$$J(\mathbf{q}_\perp) = \frac{g^2}{2m(\mathbf{q}_\perp^2 + \lambda^2)}. \tag{B70}$$

Accordingly, we define the coefficients

$$a(\mu) = (4\pi)^{-1} \int_0^\infty dz \, \phi(z, \mu)(z + \lambda^2)^{-1}. \tag{B71}$$

In order for this $a(\mu)$ to be well defined, it is assumed throughout this section that

$$\lambda > 0, \tag{B72}$$

which is a stronger condition than relations (B46). With equation (B71), the matrix element (6.22) in the forward direction is

$$\mathscr{M}_T = is(\tfrac{1}{2}g^2/m)^2 \int_{\mathscr{S}} d\mu [a(\mu)]^2 s^{g^4\mu/(8\pi^4)}. \tag{B73}$$

By equation (B26), the power c in

$$\mathscr{M}_T \sim \text{const } s^{1+c}/(\ln s)^{c'} \tag{B74}$$

is given by

$$c = \frac{g^4 \mu_0}{8\pi^4} = \frac{11g^4}{516\pi}, \tag{B75}$$

in agreement with expression (6.23).

In order to calculate c', the behavior of $a(\mu)$ for μ near μ_0 is needed. Because the continuum eigenfunctions are orthogonal, it follows from relation (B64) that

$$\int_0^\infty dz \, \phi(z, \mu)\phi(z, \mu') = N(\mu)\delta(\tau - \tau'), \tag{B76}$$

with

$$N(\mu) \sim \tfrac{1}{2}[C_1(\mu)]^2 \tag{B77}$$

for μ near μ_0. By relation (B63), a comparison of equations (B76) and (B68) shows that

$$C_1(\mu) \sim A_0^{1/4}(\mu_0 - \mu)^{-1/4}, \tag{B78}$$

and hence by relation (B67)

$$C_0(\mu) \sim A_0^{3/4}(\mu_0 - \mu)^{-1/4}, \tag{B79}$$

for μ near μ_0. Use of relation (B64) once more therefore gives the result that

$$\bar{a} = \lim_{\mu \to \mu_0} (1 - \mu/\mu_0)^{-1/4} a(\mu) \tag{B80}$$

exists and is nonzero.

Finally, the substitution of equation (B80) into equation (B73) shows that

$$c' = \tfrac{3}{2}. \tag{B81}$$

For the special case of forward scattering, expression (6.23), which gives the high-energy behavior of the tower diagrams, is just relation (B74) with c and c' given by equations (B75) and (B81), respectively.

B.5 Comments on the Case of Nonzero Momentum Transfers

So far in this appendix we have concentrated on the special case of forward scattering. What are the main ingredients used in that case? They are (1) the asymptotic behavior of the kernel for large $\mathbf{q}_\perp$ and $\mathbf{q}'_\perp$, (2) the deviation of the kernel from this asymptotic behavior when $\mathbf{q}_\perp$ and $\mathbf{q}'_\perp$ are not large, and (3) the absence of any discrete eigenvalue. For *physical* values of momentum transfers, not necessarily zero, these ingredients are all present. Therefore all the considerations in this appendix can be generalized, with a great deal more of lengthy algebra, to nonforward directions.

More precisely, a sufficient condition for expression (6.23) to hold is

$$m > 0$$

and

$$\Delta > 0 \quad \text{or} \quad \lambda > 0. \tag{B82}$$

On the other hand, when $t = -\Delta^2$ is positive, a discrete eigenvalue may emerge from the upper end of the continuous spectrum. When this happens, the asymptotic behavior of the tower diagrams in the limit of high energies is controlled by this discrete eigenvalue, and expression (6.23) no longer applies.

Reference

[1] H. Cheng and T. T. Wu, *Physical Review D* 1 (1970):2775.

Appendix C

C.1 Introduction

In this appendix we discuss through several examples the method of obtaining the high-energy behavior of various matrix elements using the Feynman parameters. Although this method is not needed in connection with the contents of this book, it serves sometimes the useful purpose of checking the results obtained by the momentum-space method.

For definiteness, we consider here only the four-point function on a mass shell with zero spin for all internal lines. Thus all propagators are of the form

$$\frac{i}{k_j^2 - m_j^2 + i\varepsilon},\tag{C1}$$

where k_j is the momentum carried by the line and $\varepsilon \to 0^+$ eventually. Let the four four-momenta be p_1, p_2, p_3, and p_4, all pointing inward. The Mandelstam variables are

$$s = (p_1 + p_2)^2, \qquad t = (p_1 + p_3)^2, \qquad u = (p_1 + p_4)^2,\tag{C2}$$

which satisfy

$$s + t + u = \sum_{j=1}^{4} M_j^2,\tag{C3}$$

with the masses M_j of the external lines given by

$$p_j^2 = M_j^2.$$

Except for an unimportant overall constant that depends on the coupling constants, the matrix element corresponding to such a Feynman diagram is

$$F = \int \prod_l d^4 k_l \prod_j \frac{1}{k_j^2 - m_j^2 + i\varepsilon},\tag{C4}$$

where $\prod_j$ is over all internal lines and $\prod_l$ is over all independent internal momenta. Thus the number of $d^4 k_l$ integrals is equal to the number of loops of the diagram. In all the examples to be discussed, the right-hand side of equation (C4) is convergent.

C.2 Feynman Parameters and Symanzik Rules [1]

One of the many useful ways of treating the right-hand side of equation (C4) employs the integral representation

$$
\prod_j \frac{1}{k_j^2 - m_j^2 + i\varepsilon}
$$

$$
= (N-1)! \int_0^\infty \left(\prod_j d\alpha_j \right) \delta \left(1 - \sum_j \alpha_j \right) \left[\sum_j \alpha_j (k_j^2 - m_j^2) + i\varepsilon \right]^{-N}, \tag{C5}
$$

where N is the number of internal lines. Thus there is a Feynman parameter α_j for each internal line, and the total number of Feynman parameters is equal to N.

With equation (C5) the momentum integrations can be carried out to give

$$
I = \int_0^\infty \left(\prod_j d\alpha_j \right) \delta \left(1 - \sum_j \alpha_j \right) [\Lambda(\alpha)]^{-2} \left[Q(s,t,u;\alpha) - \sum_j \alpha_j m_j^2 + i\varepsilon \right]^{-n}, \tag{C6}
$$

where I differs from F in equation (C4) by an overall constant factor, including the $(N-1)!$ of equation (C5); n is an integer equal to N minus twice the number of loops; and $Q(s,t,u;\alpha)$ is the extremal value of the quadratic form $\sum_j \alpha_j k_j^2$ with all α_j and the four external momenta p_i fixed. (Here α means the set of all N α_j.) Because the sum $\sum_j \alpha_j k_j^2$ is a quadratic form in the momenta, $Q(s,t,u;\alpha)$ can be expressed as

$$
Q(s,t,u;\alpha) = Q_s(\alpha)s + Q_t(\alpha)t + Q_u(\alpha)u + \sum_{i=1}^4 Q_i(\alpha)M_i^2. \tag{C7}
$$

These Q are related to the power dissipation in electric circuit theory, which is a useful connection for various purposes, such as the study of the analytic behavior of Feynman amplitudes [2]. Symanzik has given explicit rules to write Λ and these Q.

The Symanzik rules are as follows [1]:

1. Λ is given by the sum of all products of l factors of α_j such that the removal of these α_j leaves the diagram simply connected.

2. The quantity

$$
D_s = \Lambda Q_s
$$

is given by the sum of all products of $(l + 1)$ factors of α_j such that their removal leaves the diagram in the form of two simply connected pieces with p_1 and p_2 in one and p_3 and p_4 in the other piece.

3. Similar rules apply for

$$D_t = \Lambda Q_t \quad \text{and} \quad D_u = \Lambda Q_u.$$

4. The quantity

$$D_i = \Lambda Q_i$$

is given by the similar rule with p_i in one piece and the other three p in the other piece.

Examples of some of these D are given in the latter sections of this appendix.

In terms of these D, an alternative form for the I of equation (C6) is

$$I = \int_0^\infty \left(\prod_j d\alpha_j \right) \delta\left(1 - \sum_j \alpha_j \right) [\Lambda(\alpha)]^{n-2} [D(s, t, u; \alpha) + i\varepsilon]^{-n}, \tag{C8}$$

with

$$D(s, t, u; \alpha) = D_s(\alpha)s + D_t(\alpha)t + D_u(\alpha)u + D_m(\alpha), \tag{C9}$$

$$D_m(\alpha) = -\Lambda(\alpha) \sum_j \alpha_j m_j^2 + \sum_{i=1}^4 D_i(\alpha) M_i^2. \tag{C10}$$

C.3 Example 1: The Box Diagram

We begin with one of the simplest examples, the box diagram shown in figure C.1. In this and the following examples, we take all the masses, both

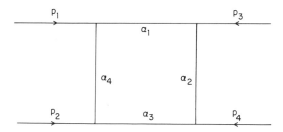

Figure C.1
A box diagram.

internal and external, to be 1. Application of the Symanzik rules gives

$$\Lambda = \alpha_1 + \alpha_2 + \alpha_3 + \alpha_4,$$

$$D_s = \alpha_1 \alpha_3, \qquad D_t = \alpha_2 \alpha_4, \qquad D_u = 0,$$

$$D_1 = \alpha_1 \alpha_4, \qquad D_2 = \alpha_3 \alpha_4, \qquad D_3 = \alpha_1 \alpha_2, \qquad D_4 = \alpha_2 \alpha_3. \qquad \text{(C11)}$$

Therefore

$$I = \int_0^\infty d\alpha_1 \, d\alpha_2 \, d\alpha_3 \, d\alpha_4 \, \delta(1 - \sum \alpha_j) \{ \alpha_1 \alpha_3 s + \alpha_2 \alpha_4 t$$

$$- [1 - (\alpha_1 + \alpha_3)(\alpha_2 + \alpha_4)] + i\varepsilon \}^{-2}. \qquad \text{(C12)}$$

The task here is to calculate the leading behavior of this I for large s with t fixed.

Mellin transformation offers a powerful method of obtaining such asymptotic behavior. Define

$$\tilde{I}(\zeta) = \int_0^\infty ds \, s^{-\zeta} I(s). \qquad \text{(C13)}$$

The point of using a Mellin transform here is that, for $\beta > -1$,

$$I(s) \sim C s^{-1} (\ln s)^\beta \quad \text{for large } s \qquad \text{(C14)}$$

if and only if

$$\tilde{I}(\zeta) \sim C \Gamma(\beta + 1) \zeta^{-1-\beta} \qquad \text{(C15)}$$

as $\zeta \to 0^+$ [3]. It is easier to study a power dependence than a logarithmic dependence.

Because

$$\int_0^\infty (As + B)^{-n} s^{-\zeta} \, ds = \frac{\pi}{\sin \pi \zeta} \binom{-\zeta}{n-1} (-1)^{n-1} A^{-1+\zeta} B^{-n+1-\zeta}, \qquad \text{(C16)}$$

the substitution of equation (C12) into equation (C13) gives

$$\tilde{I}(\zeta) = \int_0^\infty d\alpha_1 \, d\alpha_2 \, d\alpha_3 \, d\alpha_4 \, \delta(1 - \sum \alpha_j) \frac{\pi \zeta}{\sin \pi \zeta} (\alpha_1 \alpha_3)^{-1+\zeta}$$

$$\times [\alpha_2 \alpha_4 t - 1 + (\alpha_1 + \alpha_3)(\alpha_2 + \alpha_4) + i\varepsilon]^{-1-\zeta}. \qquad \text{(C17)}$$

For small ζ, both the α_1 and the α_3 integrations give a factor ζ^{-1}. Thus $\tilde{I}(\zeta)$

is of order ζ^{-2}, and all the other ζ can be replaced by 0. Because $t \leqslant 0$, the result is

$$\tilde{I}(\zeta) \sim -\zeta^{-2} \int_0^\infty d\alpha_2 \, d\alpha_4 \, \delta(1 - \alpha_2 - \alpha_4)(\alpha_2 \alpha_4 |t| + 1)^{-1}. \tag{C18}$$

Note that this simple approximation gives only the leading term; if the first correction term is needed, a more elaborate approximation is necessary. Comparison with relations (C14) and (C15) shows that

$$I(s) \sim -s^{-1}(\ln s) \int_0^1 d\alpha_2 [\alpha_2(1 - \alpha_2)|t| + 1]^{-1}. \tag{C19}$$

This is the desired result.

C.4   Example 2: The Crossed Box Diagram

It is instructive to consider briefly the diagram in figure C.2, which is merely the box diagram of figure C.1 with p_2 and p_4 exchanged, or equivalently, with s and u exchanged. Therefore the corresponding integral is, from equation (C12),

$$I' = \int_0^\infty d\alpha_1 \, d\alpha_2 \, d\alpha_3 \, d\alpha_4 \, \delta(1 - \sum \alpha_j)\{\alpha_1 \alpha_3 u + \alpha_2 \alpha_4 t$$

$$- [1 - (\alpha_1 + \alpha_3)(\alpha_2 + \alpha_4)] + i\varepsilon\}^{-2}. \tag{C20}$$

The major difference between equations (C12) and (C20) is that in equation (C20) the $i\varepsilon$ can be omitted at the beginning because the quantity in the braces is negative. Therefore I' is real, whereas the I of equation (C12) is

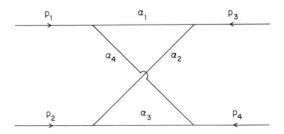

Figure C.2
A crossed box diagram.

complex. With (C20) trivially rewritten as

$$I' = \int_0^\infty d\alpha_1 \, d\alpha_2 \, d\alpha_3 \, d\alpha_4 \, \delta(1 - \sum \alpha_j) [\alpha_1 \alpha_3 (s + t - 4) - \alpha_2 \alpha_4 t + 1$$

$$- (\alpha_1 + \alpha_3)(\alpha_2 + \alpha_4)]^{-2}, \quad (C21)$$

repetition of the procedure of section C.3 gives immediately

$$I'(s) \sim s^{-1}(\ln s) \int_0^1 d\alpha_2 [\alpha_2(1 - \alpha_2)|t| + 1]^{-1} \qquad (C22)$$

for large s.

Because the right-hand sides of relations (C19) and (C22) differ only in an overall sign, we get

$$I(s) + I'(s) = o(s^{-1} \ln s) \qquad (C23)$$

for large s. The question is, Can we get the asymptotic behavior of $I(s) + I'(s)$ without extensive additional calculation? The answer is fortunately yes. Because fractional powers of ζ^{-1} cannot appear in the Mellin transform of $I'(s)$, result (C22) can be rewritten in a sharper form for large s (see also section C.10):

$$I'(s) \sim s^{-1}(\ln s) \int_0^1 d\alpha_2 [\alpha_2(1 - \alpha_2)|t| + 1]^{-1} + C_2 s^{-1}, \qquad (C24)$$

where C_2 is a real function of t. Because

$$I(s) = I'(s)|_{s \leftrightarrow u} \sim I'(s)|_{s \to s \exp(-i\pi)}, \qquad (C25)$$

it follows from relation (C24) that

$$I(s) \sim -s^{-1}(\ln s - i\pi) \int_0^1 d\alpha_2 [\alpha_2(1 - \alpha_2)|t| + 1]^{-1} - C_2 s^{-1}. \qquad (C26)$$

Because of the cancellation of C_2, the sum $I(s) + I'(s)$ is given by

$$I(s) + I'(s) \sim i\pi s^{-1} \int_0^1 d\alpha_2 [\alpha_2(1 - \alpha_2)|t| + 1]^{-1} \qquad (C27)$$

for large s. This is the desired answer. (In combining asymptotic behavior and analytic continuation, in general the possible presence of Stoke's line needs to be checked. Such complications do not appear in the high-energy asymptotic evaluation of Feynman diagrams.)

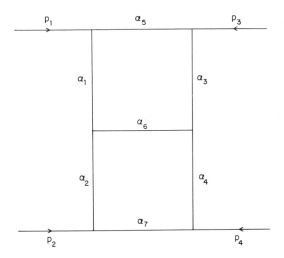

Figure C.3
A ladder diagram.

C.5 Example 3: A Ladder Diagram

As the next example, consider the ladder diagram of figure C.3, which is
the first t-channel iterate of the box diagram of figure C.1. In this case the
results of applying the Symanzik rules are

$$\Lambda = (\alpha_1 + \alpha_3 + \alpha_5 + \alpha_6)(\alpha_2 + \alpha_4 + \alpha_6 + \alpha_7) - \alpha_6^2,$$

$$D_s = \alpha_5 \alpha_6 \alpha_7,$$

$$D_t = \alpha_1 \alpha_3 (\alpha_2 + \alpha_4 + \alpha_6 + \alpha_7) + \alpha_2 \alpha_4 (\alpha_1 + \alpha_3 + \alpha_5 + \alpha_6)$$
$$\quad + \alpha_6 (\alpha_1 \alpha_4 + \alpha_2 \alpha_3),$$

$$D_u = 0,$$

and

$$D_m = -\Lambda + \alpha_5 (\alpha_1 + \alpha_3)(\alpha_2 + \alpha_4 + \alpha_6 + \alpha_7) + \alpha_7 (\alpha_2 + \alpha_4)$$
$$\quad \times (\alpha_1 + \alpha_3 + \alpha_5 + \alpha_6) + \alpha_5 \alpha_6 (\alpha_2 + \alpha_4) + \alpha_6 \alpha_7 (\alpha_1 + \alpha_3). \tag{C28}$$

In terms of these quantities, the integral I is

$$I = \int_0^\infty \left(\prod_{j=1}^7 d\alpha_j \right) \delta \left(1 - \sum_{j=1}^7 \alpha_j \right) \Lambda (D_s s + D_t t + D_m + i\varepsilon)^{-3}, \tag{C29}$$

and, by the Mellin transform (C13),

$$\tilde{I}(\zeta) = \int_0^\infty \left(\prod_{j=1}^7 d\alpha_j \right) \delta \left(1 - \sum_{j=1}^7 \alpha_j \right) \Lambda (\alpha_5 \alpha_6 \alpha_7)^{-1+\zeta} (D_t t + D_m + i\varepsilon)^{-2-\zeta}.$$

(C30)

The leading behavior for $\zeta \to 0^+$ is controlled by the α_5, α_6, and α_7 integrations. Setting $\alpha_5 = \alpha_6 = \alpha_7 = \zeta = 0$ everywhere except in the factor $(\alpha_5 \alpha_6 \alpha_7)^{1-\zeta}$, we get immediately

$$\tilde{I}(\zeta) \sim \zeta^{-3} \int_0^\infty \left(\prod_{j=1}^4 d\alpha_j \right) \delta \left(1 - \sum_{j=1}^4 \alpha_j \right) \Lambda_0 (D_{t0} |t| + \Lambda_0)^{-2},$$

(C31)

where

$$\Lambda_0 = (\alpha_1 + \alpha_3)(\alpha_2 + \alpha_4),$$

$$D_{t0} = \alpha_1 \alpha_2 \alpha_3 + \alpha_2 \alpha_3 \alpha_4 + \alpha_3 \alpha_4 \alpha_1 + \alpha_4 \alpha_1 \alpha_2.$$

(C32)

Comparison with relations (C14) and (C15) then shows that

$$I \sim \tfrac{1}{2} s^{-1} (\ln s)^2 \int_0^\infty \left(\prod_{j=1}^4 d\alpha_j \right) \delta \left(1 - \sum_{j=1}^4 \alpha_j \right) \Lambda_0 (D_{t0} |t| + \Lambda_0)^{-2}$$

(C33)

for large s. A suitable change of variables yields the much more instructive formula

$$I \sim \tfrac{1}{2} s^{-1} (\ln s)^2 \left\{ \int_0^1 d\alpha_2 [\alpha_2 (1 - \alpha_2) |t| + 1]^{-1} \right\}^2$$

(C34)

for large s. Thus the coefficient on the right-hand side of relation (C34) is essentially the square of that of relation (C19). This is precisely the relation that led to Regge poles over twenty years ago by summing the leading terms of ladder diagrams [4].

C.6   Example 4: Three-Particle Exchange

The fourth example concerns the three-particle exchange diagram of figure C.4; it is the ladder diagram of figure C.3 with p_2 and p_3 interchanged. Therefore, for this diagram, equation (C28) holds except that D_s and D_t are interchanged:

$$D_s = \text{the } D_t \text{ of equation (C28)},$$

$$D_t = \alpha_5 \alpha_6 \alpha_7.$$

(C35)

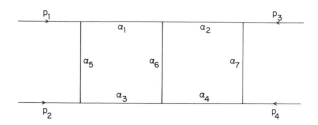

Figure C.4
A three-particle exchange diagram.

For large s, the present I, given by equation (C29) with (C35), turns out to be of the order of $s^{-2} \ln s$. Because of this s^{-2} instead of s^{-1}, it is more convenient in this case to redefine ζ so that

$$\tilde{I}(\zeta) = \int_0^\infty ds\, s^{-\zeta+1} I(s), \tag{C36}$$

instead of equation (C13). The advantage of this new ζ is that the limit of interest remains $\zeta \to 0^+$; with the old ζ, the limit would have been $\zeta \to -1^+$. With equation (C36), $\tilde{I}(\zeta)$ is given by, for $\zeta \to 0^+$,

$$\tilde{I}(\zeta) \sim \int_0^\infty \left(\prod_{j=1}^7 d\alpha_j \right) \delta\left(1 - \sum_{j=1}^7 \alpha_j\right) \Lambda D_s^{-2+\zeta} (D_t t + D_m + i\varepsilon)^{-1-\zeta}. \tag{C37}$$

The important region of contribution is the neighborhood of

$$\alpha_1 = \alpha_2 = \alpha_3 = \alpha_4 = 0. \tag{C38}$$

Let

$$\alpha_1 = \rho x, \qquad \alpha_2 = \rho(1 - x),$$
$$\alpha_3 = \rho' y, \qquad \alpha_4 = \rho'(1 - y). \tag{C39}$$

Then

$$\tilde{I}(\zeta) \sim \int_0^\infty d\rho\, d\rho'\, d\alpha_5\, d\alpha_6\, d\alpha_7 \int_0^1 dx\, dy\, \delta(1 - \rho - \rho' - \alpha_5 - \alpha_6 - \alpha_7)$$

$$\times (\rho\rho')^{-1+\zeta} \Lambda (D_t t + D_m + i\varepsilon)^{-1-\zeta} [xy(\alpha_6 + \alpha_7)$$

$$+ (1 - x)(1 - y)(\alpha_5 + \alpha_6) + \alpha_6(x + y - 2xy)]^{-2+\zeta}. \tag{C40}$$

The situation here is now quite similar to those of the previous examples: Except for the factor $(\rho\rho')^{-1+\zeta}$, it is permissible to put $\rho = \rho' = \zeta = 0$. Therefore

$$\tilde{I}(\zeta) \sim -\zeta^{-2} \int_0^\infty d\alpha_5 \, d\alpha_6 \, d\alpha_7 \int_0^1 dx \, dy \, \delta(1 - \alpha_5 - \alpha_6 - \alpha_7)$$

$$\times (\alpha_5\alpha_6 + \alpha_6\alpha_7 + \alpha_7\alpha_5)(\alpha_5\alpha_6\alpha_7|t| + \alpha_5\alpha_6 + \alpha_6\alpha_7 + \alpha_7\alpha_5)^{-1}$$

$$\times [xy(\alpha_6 + \alpha_7) + (1 - x)(1 - y)(\alpha_5 + \alpha_6) + \alpha_6(x + y - 2xy)]^{-2}. \quad (C41)$$

In this form it is straightforward to carry out the x and y integrations to give

$$I(s) \sim -s^{-2}(\ln s) \int_0^\infty d\alpha_5 \, d\alpha_6 \, d\alpha_7 \, \delta(1 - \alpha_5 - \alpha_6 - \alpha_7)$$

$$\times (\alpha_5\alpha_6\alpha_7|t| + \alpha_5\alpha_6 + \alpha_6\alpha_7 + \alpha_7\alpha_5)^{-1} \ln \frac{(\alpha_5 + \alpha_6)(\alpha_6 + \alpha_7)}{\alpha_6^2}. \quad (C42)$$

This is the desired leading term in the asymptotic behavior for large s with fixed t. Note that this leading term is purely real.

It is also worthwhile to note that, although the diagrams of figures C.3 and C.4 are simply related by the exchange of s and t, their asymptotic behavior for large s is quite different, that of figure C.4 being far more complicated.

C.7   Example 5: A Second Three-Particle Exchange Diagram

In all the examples discussed so far, the leading asymptotic behavior for large s is purely real. In this fifth example, shown in figure C.5, we present a case in which this is not so.

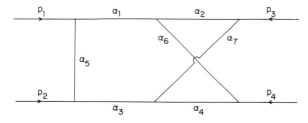

Figure C.5
Another three-particle diagram.

In this case the various quantities are

$$\Lambda = (\alpha_1 + \alpha_3 + \alpha_5)(\alpha_2 + \alpha_4 + \alpha_6 + \alpha_7) + (\alpha_2 + \alpha_7)(\alpha_4 + \alpha_6),$$

$$D_s = \alpha_1 \alpha_3 (\alpha_2 + \alpha_4 + \alpha_6 + \alpha_7) + \alpha_1 \alpha_4 \alpha_7 + \alpha_2 \alpha_3 \alpha_6,$$

$$D_t = \alpha_5 \alpha_6 \alpha_7,$$

$$D_u = \alpha_2 \alpha_4 \alpha_5,$$

and

$$D_m = -\Lambda + \alpha_5(\alpha_1 + \alpha_3)(\alpha_2 + \alpha_4 + \alpha_6 + \alpha_7) + \alpha_5(\alpha_2 \alpha_6 + \alpha_4 \alpha_7)$$

$$+ \alpha_2 \alpha_7(\alpha_1 + \alpha_3 + \alpha_4 + \alpha_5 + \alpha_6) + \alpha_4 \alpha_6(\alpha_1 + \alpha_2 + \alpha_3 + \alpha_5 + \alpha_7)$$

$$+ (\alpha_1 + \alpha_3)(\alpha_2 \alpha_4 + \alpha_6 \alpha_7). \tag{C43}$$

In terms of these quantities,

$$I = \int_0^\infty \left(\prod_{j=1}^7 d\alpha_j \right) \delta\left(1 - \sum_{j=1}^7 \alpha_j \right) \Lambda (D_s s + D_t t + D_u u + D_m + i\varepsilon)^{-3}. \tag{C44}$$

By equation (C3), the Mellin transform as defined by equation (C36) is

$$\tilde{I}(\zeta) \sim \int_0^\infty \left(\prod_{j=1}^7 d\alpha_j \right) \delta\left(1 - \sum_{j=1}^7 \alpha_j \right) \Lambda (D_s - D_u + i\varepsilon)^{-2+\zeta}$$

$$\times [(D_t - D_u)t + (D_m + 4D_u) + i\varepsilon]^{-1-\zeta}. \tag{C45}$$

Note the necessity of keeping the $i\varepsilon$ term together with $D_s - D_u$.

The remaining considerations are entirely similar to those of relation (C37) in the preceding section: The important region of contribution for small ζ is the neighborhood of equation (C38), and the change of variables (C39) remains appropriate. The analogue of relation (C41) is

$$\tilde{I}(\zeta) \sim -\zeta^2 \int_0^\infty d\alpha_5 \, d\alpha_6 \, d\alpha_7 \int_0^1 dx \, dy \, \delta(1 - \alpha_5 - \alpha_6 - \alpha_7)$$

$$\times (\alpha_5 \alpha_6 + \alpha_6 \alpha_7 + \alpha_7 \alpha_5)(\alpha_5 \alpha_6 \alpha_7 |t| + \alpha_5 \alpha_6 + \alpha_6 \alpha_7 + \alpha_7 \alpha_5)^{-1}$$

$$\times [xy(\alpha_6 + \alpha_7) + x(1 - y)\alpha_7 + (1 - x)y\alpha_6 - (1 - x)(1 - y)\alpha_5 + i\varepsilon]^{-2} \tag{C46}$$

for $\zeta \to 0^+$. Carrying out the x and y integrations gives

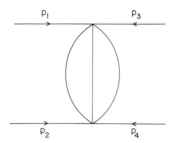

Figure C.6
The Feynman diagram obtained from figures C.4 and C.5 by shrinking lines 1, 2, 3, and 4.

$$I(s) \sim s^{-2}(\ln s) \int_0^\infty d\alpha_5 d\alpha_6 d\alpha_7 \, \delta(1 - \alpha_5 - \alpha_6 - \alpha_7)$$

$$\times (\alpha_5 \alpha_6 \alpha_7 |t| + \alpha_5 \alpha_6 + \alpha_6 \alpha_7 + \alpha_7 \alpha_5)^{-1} \left(\ln \frac{\alpha_6 + \alpha_7}{\alpha_7} + i\pi \right), \quad (C47)$$

where an additional term $\ln(\alpha_5/\alpha_7)$ inside the last parentheses has been dropped because of the invariance of the rest of the integrand under the exchange of α_5 and α_7.

This result (C47) shows that the leading asymptotic behavior is complex in this case. Furthermore, the real part differs from the result of the preceding section only by a factor of $-\frac{1}{2}$. On the other hand, the imaginary part of the right-hand side of relation (C47) is proportional to the Feynman integral for the diagram with lines 1, 2, 3, and 4 shrunk, that is, the diagram of figure C.6. Generalizations of such relations also hold for more complicated multiparticle exchange diagrams.

C.8   Comments on Three-Particle Exchange Diagrams

In section C.4 we discussed the sum of the two two-particle exchange diagrams of figures C.1 and C.2. Here we give the corresponding results for the six three-particle exchange diagrams shown in figure C.7. Of course, figures C.7a and C.7b are the same as figures C.4 and C.5, respectively. Figures C.7d through C.7f are obtained from figures C.7a through C.7c by the interchange of p_2 and p_4 or, equivalently, $s \leftrightarrow u$.
    Define

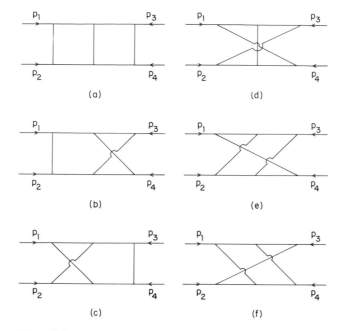

Figure C.7
The six three-particle exchange diagrams.

$$\begin{bmatrix} C_{31}(t) \\ C_{30}(t) \end{bmatrix} = \int_0^\infty \frac{d\alpha_5\, d\alpha_6\, d\alpha_7\, \delta(1 - \alpha_5 - \alpha_6 - \alpha_7)}{\alpha_5 \alpha_6 \alpha_7 |t| + \alpha_5 \alpha_6 + \alpha_6 \alpha_7 + \alpha_7 \alpha_5} \begin{bmatrix} \ln \dfrac{\alpha_6 + \alpha_7}{\alpha_7} \\ \pi \end{bmatrix}. \qquad \text{(C48)}$$

In terms of these two functions of t, the leading asymptotic behavior of the six integrals for large s is given by

(a) $-2C_{31}s^{-2}\ln(se^{-i\pi})$,

(b) $(C_{31} + iC_{30})s^{-2}\ln(se^{-i\pi})$,

(c) $(C_{31} + iC_{30})s^{-2}\ln(se^{-i\pi})$,

(d) $-2C_{31}s^{-2}\ln s$,

(e) $(C_{31} - iC_{30})s^{-2}\ln s$,

(f) $(C_{31} - iC_{30})s^{-2}\ln s$.

Written in this way, these formulas correctly give the leading asymptotic

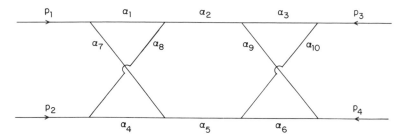

Figure C.8
A four-particle exchange diagram.

behavior of the sum of the various matrix elements (see the discussion of section C.4). In particular, with C_2-type terms canceled,

(a) + (b) + (c) $2iC_{30}s^{-2}\ln(se^{-i\pi})$,

(d) + (e) + (f) $-2iC_{30}s^{-2}\ln s$,

(a) + (b) + (c) + (d) + (e) + (f) $2\pi C_{30}s^{-2}$.

Thus there is no factor of $\ln s$ in the sum of all six matrix elements for the diagrams of figure C.7.

C.9   Example 6: A Four-Particle Exchange Diagram

As the final example, we discuss the four-particle exchange diagram of figure C.8. This example illustrates a new feature: The leading contributions for large s come from not one but rather two distinct regions.

This is perhaps to be expected because, unlike all the earlier examples, this diagram has a symmetry number of 2. More precisely, the diagram is invariant under the following exchange of internal lines as labeled by the indexes of the Feynman parameters: $1 \leftrightarrow 7$, $2 \leftrightarrow 5$, $3 \leftrightarrow 10$, $4 \leftrightarrow 8$, and $6 \leftrightarrow 9$. On the basis of the experience with the preceding three sections, the two regions of contribution to the leading terms for large s are, respectively, the neighborhoods of

$$\alpha_1 = \alpha_2 = \alpha_3 = \alpha_4 = \alpha_5 = \alpha_6 = 0, \tag{C49}$$

$$\alpha_2 = \alpha_5 = \alpha_7 = \alpha_8 = \alpha_9 = \alpha_{10} = 0. \tag{C50}$$

Because the sum of all ten α is 1, these two neighborhoods are disconnected. By the symmetry of the diagram, the two contributions are equal. Therefore it is sufficient to consider the neighborhood of the point defined in equation (C49) and multiply the result by 2. For this diagram the quantities are

$$\Lambda = (\alpha_2 + \alpha_5)(\alpha_1 + \alpha_4 + \alpha_7 + \alpha_8)(\alpha_3 + \alpha_6 + \alpha_9 + \alpha_{10})$$

$$+ (\alpha_1 + \alpha_4 + \alpha_7 + \alpha_8)(\alpha_3 + \alpha_{10})(\alpha_6 + \alpha_9)$$

$$+ (\alpha_3 + \alpha_6 + \alpha_9 + \alpha_{10})(\alpha_1 + \alpha_7)(\alpha_4 + \alpha_8),$$

$$D_s = \alpha_2\alpha_5(\alpha_1 + \alpha_4 + \alpha_7 + \alpha_8)(\alpha_3 + \alpha_6 + \alpha_9 + \alpha_{10})$$

$$+ (\alpha_1 + \alpha_4 + \alpha_7 + \alpha_8)(\alpha_2\alpha_6\alpha_{10} + \alpha_3\alpha_5\alpha_9)$$

$$+ (\alpha_3 + \alpha_6 + \alpha_9 + \alpha_{10})(\alpha_1\alpha_5\alpha_8 + \alpha_2\alpha_4\alpha_7)$$

$$+ \alpha_1\alpha_6\alpha_8\alpha_{10} + \alpha_3\alpha_4\alpha_7\alpha_9,$$

$$D_t = \alpha_1\alpha_3\alpha_4\alpha_6 + \alpha_7\alpha_8\alpha_9\alpha_{10},$$

$$D_u = \alpha_1\alpha_4\alpha_9\alpha_{10} + \alpha_3\alpha_6\alpha_7\alpha_8,$$

$$D_m = -\Lambda + (\alpha_1 + \alpha_4 + \alpha_7 + \alpha_8)[(\alpha_2 + \alpha_5)(\alpha_3 + \alpha_9)(\alpha_6 + \alpha_{10})$$

$$+ \alpha_3\alpha_{10}(\alpha_6 + \alpha_9) + \alpha_6\alpha_9(\alpha_3 + \alpha_{10})]$$

$$+ (\alpha_3 + \alpha_6 + \alpha_9 + \alpha_{10})[(\alpha_2 + \alpha_5)(\alpha_1 + \alpha_8)(\alpha_4 + \alpha_7)$$

$$+ \alpha_1\alpha_7(\alpha_4 + \alpha_8) + \alpha_4\alpha_8(\alpha_1 + \alpha_7)]$$

$$+ (\alpha_1 + \alpha_7)(\alpha_4 + \alpha_8)(\alpha_3\alpha_{10} + \alpha_6\alpha_9)$$

$$+ (\alpha_3 + \alpha_{10})(\alpha_6 + \alpha_9)(\alpha_1\alpha_7 + \alpha_4\alpha_8)$$

$$+ (\alpha_1\alpha_4 + \alpha_7\alpha_8)(\alpha_3\alpha_9 + \alpha_6\alpha_{10}) + (\alpha_1\alpha_8 + \alpha_4\alpha_7)(\alpha_3\alpha_6 + \alpha_9\alpha_{10}).$$

$$\text{(C51)}$$

In terms of these quantities, I is given by

$$I = \int_0^\infty \left(\prod_{j=1}^{10} d\alpha_j \right) \delta\left(1 - \sum_{j=1}^{10} \alpha_j \right) \Lambda^2 (D_s s + D_t t + D_u u + D_m + i\varepsilon)^{-4}. \quad \text{(C52)}$$

Similar to equations (C13) and (C36), the appropriate Mellin transform for the present case is

$$\tilde{I}(\zeta) = \int_0^\infty ds \, s^{-\zeta+2} I(s). \quad \text{(C53)}$$

By equations (C3) and (C16), this $\tilde{I}(\zeta)$ is given by

$$\tilde{I}(\zeta) = -\frac{\pi}{\sin \pi \zeta}\binom{2-\zeta}{3}\int_0^\infty \left(\prod_{j=1}^{10} d\alpha_j\right)\delta\left(1 - \sum_{j=1}^{10}\alpha_j\right)\Lambda^2$$

$$\times (D_s - D_u + i\varepsilon)^{-3+\zeta}[(D_t - D_u)t + (D_m + 4D_u) + i\varepsilon]^{-1-\zeta}. \quad (C54)$$

In the neighborhood of the point defined in equation (C49), it is convenient to use the variables

$$\rho = \alpha_1 + \alpha_2 + \alpha_3,$$

$$\rho' = \alpha_4 + \alpha_5 + \alpha_6,$$

$$\alpha_1 = \rho x_1, \qquad \alpha_2 = \rho x_2, \qquad \alpha_3 = \rho x_3,$$

$$\alpha_4 = \rho' y_1, \qquad \alpha_5 = \rho' y_2, \qquad \alpha_6 = \rho' y_3, \qquad (C55)$$

analogous to equations (C39). By keeping only the leading terms in ρ and ρ', we see that the following approximations hold:

$$\Lambda \sim \Lambda_0, \qquad D_s \sim \rho\rho' D_{s0}, \qquad D_t \sim D_{t0},$$

$$D_u \sim \rho\rho' D_{u0}, \qquad D_m \sim -\Lambda_0, \qquad (C56)$$

where

$$\Lambda_0 = (\alpha_7 + \alpha_8)\alpha_9\alpha_{10} + (\alpha_9 + \alpha_{10})\alpha_7\alpha_8,$$

$$D_{s0} = (x_2 + x_3)(y_1 + y_2)\alpha_7\alpha_9 + (x_1 + x_2)(y_2 + y_3)\alpha_8\alpha_{10}$$

$$+ x_2\alpha_7\alpha_{10} + y_2\alpha_8\alpha_9,$$

$$D_{t0} = \alpha_7\alpha_8\alpha_9\alpha_{10},$$

$$D_{u0} = x_1 y_1 \alpha_9\alpha_{10} + x_3 y_3 \alpha_7\alpha_8. \qquad (C57)$$

With relations (C56) and by carrying out explicitly the ρ and ρ' integrations, we find that the leading term for the $\tilde{I}(\zeta)$ of equation (C54) for small $\zeta > 0$ is

$$\tilde{I}(\zeta) \sim \tfrac{2}{3}\zeta^{-2}\int_0^\infty d\alpha_7\,d\alpha_8\,d\alpha_9\,d\alpha_{10}\,dx_1\,dx_2\,dx_3\,dy_1\,dy_2\,dy_3$$

$$\times \delta(1 - \alpha_7 - \alpha_8 - \alpha_9 - \alpha_{10})\delta(1 - x_1 - x_2 - x_3)$$

$$\times \delta(1 - y_1 - y_2 - y_3)\Lambda_0^2(D_{s0} - D_{u0} + i\varepsilon)^{-3+\zeta}(D_{t0}t - \Lambda_0)^{-1}, \quad (C58)$$

where a factor of 2 has been included, as explained in the discussion following equation (C50).

The important point is that there is a further factor of ζ^{-1} resulting from $(D_{s0} - D_{u0} + i\varepsilon)^{-3+\zeta}$. Thus $\tilde{I}(\zeta) = O(\zeta^{-3})$ for small ζ. This is seen by writing $D_{s0} - D_{u0}$ explicitly from equations (C57):

$$D_{s0} - D_{u0}$$

$$= -(x_1\alpha_{10} - x_3\alpha_7)(y_1\alpha_9 - y_3\alpha_8) + x_2(y_1\alpha_7\alpha_9 + y_3\alpha_8\alpha_{10} + \alpha_7\alpha_{10})$$

$$+ y_2(x_3\alpha_7\alpha_9 + x_1\alpha_8\alpha_{10} + \alpha_8\alpha_9) + x_2y_2(\alpha_7\alpha_9 + \alpha_8\alpha_{10}). \tag{C59}$$

The leading contribution comes from the neighborhood of

$$x_2 = 0, \qquad y_2 = 0,$$

$$x_1\alpha_{10} - x_3\alpha_7 = 0, \qquad y_1\alpha_9 - y_3\alpha_8 = 0, \tag{C60}$$

which imply

$$x_1 = \alpha_7(\alpha_7 + \alpha_{10})^{-1}, \qquad x_3 = \alpha_{10}(\alpha_7 + \alpha_{10})^{-1},$$

$$y_1 = \alpha_8(\alpha_8 + \alpha_9)^{-1}, \qquad y_3 = \alpha_9(\alpha_8 + \alpha_9)^{-1}. \tag{C61}$$

After the x_2 and y_2 integrations are carried out, the $\tilde{I}(\zeta)$ of relation (C58) becomes

$$\tilde{I}(\zeta) \sim \tfrac{1}{3}\zeta^{-2} \int_0^\infty d\alpha_7 \, d\alpha_8 \, d\alpha_9 \, d\alpha_{10} \, dx_1 \, dx_3 \, dy_1 \, dy_3 \, \delta(1 - \alpha_7 - \alpha_8 - \alpha_9 - \alpha_{10})$$

$$\times \, \delta(1 - x_1 - x_3)\delta(1 - y_1 - y_3)(\alpha_7 + \alpha_{10})(\alpha_8 + \alpha_9)$$

$$\times \, [-(x_1\alpha_{10} - x_3\alpha_7)(y_1\alpha_9 - y_3\alpha_8) + i\varepsilon]^{-1+\zeta}(D_{t0}t - \Lambda_0)^{-1}. \tag{C62}$$

Consider the integral

$$I_0 = \int_{-1}^1 dx \, dy(-xy + i\varepsilon)^{-1+\zeta}. \tag{C63}$$

Because the integrand is invariant under $x \to -x$, $y \to -y$, it can be evaluated explicitly:

$$I_0 = 2 \int_0^1 dx \int_{-1}^1 dy(xy + i\varepsilon)^{-1+\zeta}$$

$$= 2 \int_0^1 dx \int_{-1}^1 dy[x(y + i\varepsilon)]^{-1+\zeta}$$

$$\sim -2\pi i\zeta^{-1}. \tag{C64}$$

Therefore

$$\tilde{I}(\zeta) \sim \tfrac{2}{3}\pi i \zeta^{-3} \int_0^\infty \left(\prod_{j=7}^{10} d\alpha_j \right) \delta \left(1 - \sum_{j=7}^{10} \alpha_j \right) (-D_{t_0} t + \Lambda_0)^{-1}. \qquad \text{(C65)}$$

Contributions of the type (C64) are related to the so-called pinch singularity [5], as distinct from the endpoint singularities that appear in all the previous examples.

Explicitly, the asymptotic behavior of the I of equation (C52) for the diagram in figure C.8 is, as a consequence of relations (C53) and (C65), given by

$$I \sim \tfrac{1}{3}i\pi s^{-3}(\ln s)^2 \int_0^\infty d\alpha_7 \, d\alpha_8 \, d\alpha_9 \, d\alpha_{10} \, \delta(1 - \alpha_7 - \alpha_8 - \alpha_9 - \alpha_{10})$$

$$\times (-\alpha_7 \alpha_8 \alpha_9 \alpha_{10} t + \alpha_8 \alpha_9 \alpha_{10} + \alpha_7 \alpha_9 \alpha_{10} + \alpha_7 \alpha_8 \alpha_{10} + \alpha_7 \alpha_8 \alpha_9)^{-1}. \qquad \text{(C66)}$$

C.10   Higher-Order Terms

So far in this appendix we have discussed only the leading high-energy behavior of the contributions from one Feynman diagram or a small number of Feynman diagrams. In this section we consider briefly the nonleading contributions.

The key point in the calculation of higher-order terms is the following observation: Consider the integrand in equation (C8) with the $\delta(1 - \sum_j \alpha_j)$ deleted, that is,

$$\left(\prod_j d\alpha_j \right) [\Lambda(\alpha)]^{n-2} [D(s, t, u; \alpha) + i\varepsilon]^{-n},$$

where $n = N - 2l$, with l the number of independent loops. Because $\Lambda(\alpha)$ and $D(s, t, u; \alpha)$ are of orders l and $l + 1$, respectively, this expression is homogeneous in α of order 0, because

$$N + (n - 2)l - n(l + 1) = 0.$$

Because of this property, the factor $\delta(1 - \sum_j \alpha_j)$ can be replaced without any approximation by $\delta(1 - \sum_j' \alpha_j)$, where the sum $\sum_j'$ is over any nonempty subset of α_j. In other words, equation (C8) can be written alternatively in the form

$$I = \int_0^\infty \left(\prod_j d\alpha_j \right) \delta \left(1 - \sum_j{}' \alpha_j \right) [\Lambda(\alpha)]^{n-2} [D(s,t,u;\alpha) + i\varepsilon]^{-n}. \tag{C67}$$

Higher-order contributions are obtained by subtracting the leading terms from the exact expression. The usefulness of equation (C67) therefore stems from the possibility of matching the arguments of the δ functions in the exact expression and the leading contribution. Let us illustrate this point with the first example shown in figure C.1. In this example the leading term, as given by relation (C18), contains the factor $\delta(1 - \alpha_2 - \alpha_4)$. Accordingly, in order to get the next-to-leading-order terms near $\zeta = 0$, equation (C17) is modified:

$$\tilde{I}(\zeta) = \int_0^\infty d\alpha_1\, d\alpha_2\, d\alpha_3\, d\alpha_4\, \delta(1 - \alpha_2 - \alpha_4) \frac{\pi\zeta}{\sin\pi\zeta} (\alpha_1\alpha_3)^{-1+\zeta}$$

$$\times\, [\alpha_2\alpha_4 t - (\alpha_1 + \alpha_2 + \alpha_3 + \alpha_4)^2 + (\alpha_1 + \alpha_3)(\alpha_2 + \alpha_4) + i\varepsilon]^{-1-\zeta}. \tag{C68}$$

For small positive ζ, $\tilde{I}(\zeta)$ is of the form

$$\tilde{I}(\zeta) = A_0(t)\zeta^{-2} + A_1(t)\zeta^{-1} + O(1). \tag{C69}$$

Therefore, by relations (C13) through (C15),

$$I(s) \sim s^{-1}[A_0(t)\ln s + A_1(t)] \tag{C70}$$

for $s \to \infty$ with fixed t. The function $A_0(t)$ was given in relation (C19); here we proceed to obtain $A_1(t)$.

Using the expressions (C68) and (C18), we can express $A_1(t)$ as follows:

$$A_1(t) = \lim_{\zeta\to 0^+} \zeta \int_0^1 d\alpha_2\, d\alpha_4\, \delta(1 - \alpha_2 - \alpha_4) \left\{ \int_0^\infty d\alpha_1\, d\alpha_3 (\alpha_1\alpha_3)^{-1+\zeta} \right.$$

$$\times\, [\alpha_2\alpha_4 t - (\alpha_1 + \alpha_2 + \alpha_3 + \alpha_4)^2 + (\alpha_1 + \alpha_3)(\alpha_2 + \alpha_4) + i\varepsilon]^{-1+\zeta}$$

$$\left. -\int_0^1 d\alpha_1\, d\alpha_3 (\alpha_1\alpha_3)^{-1+\zeta} [\alpha_2\alpha_4 t - (\alpha_1 + \alpha_3)^2 + i\varepsilon]^{-1} \right\}. \tag{C71}$$

The contributions to $A_1(t)$ come from the regions where α_1 or α_3 (but not both) is small. By splitting the integrations over α_1 and α_3 into the two ranges $(0, 1)$ and $(1, \infty)$, we can readily reduce equation (C71) to

$$A_1(t) = \lim_{\zeta \to 0^+} \zeta \int_0^1 d\alpha_2\, d\alpha_4\, \delta(1 - \alpha_2 - \alpha_4) \left(2 \int_0^1 d\alpha_1 \int_1^\infty d\alpha_3 (\alpha_1 \alpha_3)^{-1+\zeta} \right.$$

$$\times \left[\alpha_2 \alpha_4 t - (\alpha_1 + \alpha_2 + \alpha_3 + \alpha_4)^2 + (\alpha_1 + \alpha_3)(\alpha_2 + \alpha_4) + i\varepsilon \right]^{-1+\zeta}$$

$$- \int_0^1 d\alpha_1\, d\alpha_3 (\alpha_1 \alpha_3)^{-1+\zeta} [\alpha_2 \alpha_4 t - (\alpha_1 + \alpha_2 + \alpha_3 + \alpha_4)^2$$

$$+ (\alpha_1 + \alpha_3)(\alpha_2 + \alpha_4) + i\varepsilon]^{-1} \zeta \ln[\alpha_2 \alpha_4 t - (\alpha_1 + \alpha_2 + \alpha_3 + \alpha_4)^2$$

$$+ (\alpha_1 + \alpha_3)(\alpha_2 + \alpha_4) + i\varepsilon]$$

$$+ \int_0^1 d\alpha_1\, d\alpha_3 (\alpha_1 \alpha_3)^{-1+\zeta} \{ \alpha_2 \alpha_4 t - (\alpha_1 + \alpha_2 + \alpha_3 + \alpha_4)^2$$

$$\left. + (\alpha_1 + \alpha_3)(\alpha_2 + \alpha_4) + i\varepsilon]^{-1} - [\alpha_2 \alpha_4 t - (\alpha_1 + \alpha_3)^2 + i\varepsilon]^{-1} \} \right). \quad \text{(C72)}$$

It is straightforward to evaluate these three pieces separately to obtain, for $t < 0$,

$$A_1(t) = - \int_0^1 d\alpha_2\, d\alpha_4\, \delta(1 - \alpha_2 - \alpha_4) \left\{ 2 \int_1^\infty d\alpha_3\, \alpha_3^{-1} \right.$$

$$\times [\alpha_2 \alpha_4 |t| + (1 + \alpha_3 + \alpha_3^2)]^{-1} - (\alpha_2 \alpha_4 |t| + 1)^{-1} [\ln(\alpha_2 \alpha_4 |t| + 1) + i\pi]$$

$$\left. - 2 \int_0^1 d\alpha_3 (1 + \alpha_3) [\alpha_2 \alpha_4 |t| + (1 + \alpha_3 + \alpha_3^2)]^{-1} [\alpha_2 \alpha_4 |t| + 1]^{-1} \right\}. \quad \text{(C73)}$$

This is the desired result. Note that the imaginary part here is precisely the expression on the right-hand side of relation (C27).

We conclude with a brief remark on further correction terms to relation (C70). The complete asymptotic series is of the form

$$I(s) \sim \sum_{n=0}^\infty s^{-n-1} [A_{2n}(t) \ln s + A_{2n+1}(t)]. \quad \text{(C74)}$$

It is in the computation of the further $A_n(t)$ that the method of Mellin transform shows its tremendous power. The functions $A_{2n}(t)$ and $A_{2n+1}(t)$ are obtained from the behavior of the analytic continuation of $\tilde{I}(\zeta)$ to the vicinity of $\zeta = -n$. After the analytic continuation, which can be explicitly carried out through integration by parts, for example, $A_{2n}(t)$ and $A_{2n+1}(t)$ are obtained in a way similar to $A_0(t)$ and $A_1(t)$, as discussed in section C.3.

References

[1] K. Symanzik, *Progress of Theoretical Physics* 20 (1958): 690.

[2] T. T. Wu, *Physical Review* 123 (1961): 678, 689.

[3] J. D. Bjorken and T. T. Wu, *Physical Review* 130 (1963): 2566.

[4] B. W. Lee and R. F. Sawyer, *Physical Review* 127 (1962): 2266; J. C. Polkinghorne, *Journal of Mathematical Physics* 4 (1963): 503; P. G. Federbush and M. T. Grisaru, *Annals of Physics* 22 (1963): 263, 299.

[5] L. D. Landau, *Nuclear Physics* 13 (1959): 181; R. J. Eden, *Physical Review* 121 (1961): 1567; P. V. Landshoff, J. C. Polkinghorne, and J. C. Taylor, *Nuovo Cimento* 19 (1961): 939; R. J. Eden, P. V. Landshoff, D. I. Olive, and J. C. Polkinghorne, *The Analytic S-Matrix* (Cambridge: Cambridge University Press, 1966).

Collaborations

Chapter 1

[12] TASSO Collaboration: R. Brandelik, W. Braunschweig, K. Gather, V. Kadansky, K. Lübelsmeyer, P. Mättig, H.-U. Martyn, G. Peise, J. Rimkus, H. G. Sander, D. Schmitz, A. Schultz von Dratzig, D. Trines, W. Wallraff, H. Boerner, H. M. Fischer, H. Hartmann, E. Hilger, W. Hillen, G. Knop, W. Korbach, P. Leu, B. Löhr, F. Roth, W. Rühmer, R. Wedemeyer, N. Wermes, M. Wollstadt, R. Bühring, R. Fohrmann, D. Heyland, H. Hultschig, P. Joos, W. Koch, U. Kötz, H. Kowalski, A. Ladage, D. Lüke, H. L. Lynch, G. Mikenberg, D. Notz, J. Pyrlik, R. Riethmüller, M. Schliwa, P. Söding, B. H. Wiik, G. Wolf, M. Holder, G. Poelz, J. Ringel, O. Römer, R. Rüsch, P. Schmüser, D. M. Binnie, P. J. Dornan, N. A. Downie, D. A. Garbutt, W. G. Jones, S. L. Lloyd, D. Pandoulas, A. Pevsner, J. Sedgbeer, S. Yarker, C. Youngman, R. J. Barlow, R. J. Cashmore, J. Illingworth, M. Ogg, G. L. Salmon, K. W. Bell, W. Chinowsky, B. Foster, J. C. Hart, J. Proudfoot, D. R. Quarrie, D. H. Saxon, P. L. Woodworth, Y. Eisenberg, U. Karshon, E. Kogan, D. Revel, E. Ronat, A. Shapira, J. Freeman, P. Lecomte, T. Meyer, S. L. Wu, and G. Zobernig, *Physics Letters B* 86 (1979): 243.

[12] MARK J Collaboration: D. P. Barber, U. Becker, H. Benda, A. Boehm, J. G. Branson, J. Bron, D. Buikman, J. Burger, C. C. Chang, H. S. Chen, M. Chen, C. P. Cheng, Y. S. Chu, R. Clare, P. Duinker, G. Y. Fang, H. Fesefeldt, D. Fong, M. Fukushima, J. C. Guo, A. Hariri, G. Herten, M. C. Ho, H. K. Hsu, T. T. Hsu, R. W. Kadel, W. Krenz, J. Li, Q. Z. Li, M. Lu, D. Luckey, D. A. Ma, C. M. Ma, G. G. G. Massaro, T. Matsuda, H. Newman, J. Paradiso, F. P. Poschmann, J. P. Revol, M. Rohde, H. Rykaczewski, K. Sinram, H. W. Tang, L. G. Tang, S. C. C. Ting, K. L. Tung, F. Vannucci, X. R. Wang, P. S. Wei, M. White, G. H. Wu, T. W. Wu, J. P. Xi, P. C. Yang, X. H. Yu, N. L. Zhang, and R. Y. Zhu, *Physical Review Letters* 43 (1979): 830.

[12] PLUTO Collaboration: C. Berger, H. Genzel, R. Grigull, W. Lackas, F. Raupach, A. Klovning, E. Lillestöl, E. Lillethun, J. A. Skard, H. Ackermann, G. Alexander, F. Barreiro, J. Bürger, L. Criegee, H. C. Dehne, R. Devenish, A. Eskreys, G. Flügge, G. Franke, W. Gabriel, C. Gerke, G. Knies, E. Lehmann, H. D. Mertiens, K. H. Pape, H. D. Reich, B. Stella, T. N. Ranga Swamy, U. Timm, W. Wagner, P. Waloschek, G. G. Winter, W. Zimmermann, O. Achterberg, V. Blobel, L. Boesten, H. Kapitza, B. Koppitz, W. Lührsen, R. Maschuw, R. van Staa, H. Spitzer, C. Y. Chang, R. G. Glasser, R. G. Kellogg, K. H. Lau, R. Sechi-Zorn, A. Skuja, G. Welch, G. T. Zorn, A. Bäcker, S. Brandt, K. Derikum, A. Diekmann, C. Grupen, H. J. Meyer, B. Neumann, M. Rost, G. Zech, T. Azemoon, H. J. Daum, H. Meyer, O. Meyer, M. Rössler, D. Schmidt, and K. Wacker, *Physics Letters B* 86 (1979): 418.

[12] JADE Collaboration: W. Bartel, T. Canzler, D. Cords, P. Dittmann, R. Eichler, R. Felst, D. Haidt, S. Kawabata, H. Krehbiel, B. Naroska, L. H. O'Neill, J. Olsson, P. Steffen, W. L. Yen, E. Elsen, M. Helm, A. Petersen, P. Warming, G. Weber, H. Drumm, J. Heintze, G. Heinzelmann, R. D. Heuer, J. von Krogh, P. Lennert, H. Matsumura, T. Nozaki, H. Rieseberg, A. Wagner, D. C. Darvill, F. Foster, G. Hughes, H. Wriedt, J. Allison, J. Armitage, I. Duerdoth, J. Hassard, F. Loebinger, B. King, A. Macbeth, H. Mills, P. G. Murphy, H. Prosper, K. Stephens, D. Clarke, M. C. Goddard, R. Hedgecock, R. Marshall, G. F. Pearce, M. Imori, T. Kobayashi, S. Komamiya, M. Koshiba, M. Minowa, S. Orito, A. Sato, T. Suda, H. Takeda, Y. Totsuka, Y. Watanabe, S. Yamada, and C. Yanagisawa, *Physics Letters B* 91 (1980): 142.

[15] UA1 Collaboration: G. Arnison, A. Astbury, B. Aubert, C. Bacci, G. Bauer, A. Bézaguet, R. Böck, T. J. V. Bowcock, M. Calvetti, T. Carroll, P. Catz, P. Cennini, S. Centro, F. Ceradini, S. Cittolin, D. Cline, C. Cochet, J. Colas, M. Corden, D. Dallman. D. Dau, M. DeBeer, M. Della Negra, M. Demoulin, D. Denegri, A. Di Ciaccio, D. DiBitonto, L. Dobrzynski, J. D. Dowell, M. Edwards, K. Eggert, E. Eisenhandler, N. Ellis, P. Erhard, H. Faissner, M. Fincke, G. Fontaine, R. Frey, R. Frühwirth, J. Garvey, S. Geer, C. Ghesquière, P. Ghez, K. L. Giboni, W. R. Gibson, Y. Giraud-Héraud, A. Givernaud, A. Gonidec, G. Grayer, P. Gutierrez, T. Hansl-Kozanecka, W. J. Haynes, L. O. Hertzberger, C. Hodges, D. Hoffmann, H. Hoffmann,

D. J. Holthuizen, R. J. Homer, A. Honma, W. Jank, G. Jorat, P. I. P. Kalmus, V. Karimäki, R. Keeler, I. Kenyon, A. Kernan, R. Kinnunen, H. Kowalski, W. Kozanecki, D. Kryn, F. Lacava, J.-P. Laugier, J.-P. Lees, H. Lehmann, R. Leuchs, A. Lévêque, D. Linglin, E. Locci, M. Loret, J.-J. Malosse, T. Markiewicz, G. Maurin, T. McMahon, J.-P. Mendiburu, M.-N. Minard, M. Mohammadi, M. Moricca, K. Morgan, H. Muirhead, F. Muller, A. K. Nandi, L. Naumann, A. Norton, A. Orkin-Lecourtois, L. Paoluzi, F. Pauss, G. Petrucci, G. Piano Mortari, E. Pietarinen, M. Pimiä, A. Placci, J. P. Porte, E. Radermacher, J. Ransdell, H. Reithler, J.-P. Revol, J. Rich, M. Rijssenbeek, C. Roberts, J. Rohlf, P. Rossi, C. Rubbia, B. Sadoulet, G. Sajot, G. Salvi, G. Salvini, J. Sass, J. Saudraix, A. Savoy-Navarro, D. Schinzel, W. Scott, T. P. Shah, M. Spiro, J. Strauss, J. Streets, K. Sumorok, F. Szoncso, D. Smith, C. Tao, G. Thompson, J. Timmer, E. Tscheslog, J. Tuominiemi, S. Van der Meer, B. Van Eijk, J.-P. Vialle, J. Vrana, V. Vuillemin, H. D. Wahl, P. Watkins, J. Wilson, C. Wulz, G. Y. Xie, M. Yvert, and E. Zurfluh, *Physics Letters B* 122 (1983): 103, 126 (1983): 398.

[15] UA2 Collaboration: P. Bagnaia, M. Banner, R. Battiston, P. Bloch, F. Bonaudi, K. Borer, M. Borghini, J.-C. Chollet, A. G. Clark, C. Conta, P. Darriulat, L. Di Lella, J. Dines-Hansen, P.-A. Dorsaz, L. Fayard, M. Fraternali, D. Froidevaux, G. Fumagalli, J.-M. Galliard, O. Gildemeister, V. G. Goggi, H. Grote, B. Hahn, H. Hänni, J. R. Hansen, P. Hansen, T. Himel, V. Hungerbühler, P. Jenni, O. Kofoed-Hansen, E. Lançon, M. Livan, S. Loucatos, B. Madsen, P. Mani, B. Mansoulié, G. C. Mantovani, L. Mapelli, B. Merkel, M. Mermikides, R. Møllerud, B. Nilsson, C. Onions, G. Parrour, F. Pastore, H. Plothow-Besch, M. Polverel, J.-P. Repellin, A. Rimoldi, A. Rothenberg, A. Roussarie, G. Sauvage, J. Schacher, J. L. Siegrist, H. M. Steiner, G. Stimpfl, F. Stocker, J. Teiger, V. Vercesi, A. R. Weidberg, H. Zaccone, J. A. Zakrzewski, and W. Zeller, *Physics Letters B* 122 (1983): 476, 129 (1983): 130.

Chapter 8

[4] UA4 Collaboration: R. Battiston, M. Bozzo, P. L. Braccini, F. Carbonara, R. Carrara, R. Castaldi, F. Cervelli, G. Chiefari, E. Drago, M. Haguenauer, B. Koene, L. Linssen, G. Matthiae, L. Merola, M. Napolitano, V. Palladino, G. Sanguinetti, G. Sciacca, G. Sette, R. van Swol, J. Timmermans, C. Vannini, J. Velasco, and F. Visco, *Physics Letters B* 115 (1982): 333, 117 (1982): 126, 127 (1983): 472.

[5] UA1 Collaboration: G. Arnison, A. Astbury, B. Aubert, C. Bacci, R. Bernabei, A. Bézaguet, R. Böck, T. J. V. Bowcock, M. Calvetti, T. Carroll, P. Catz, S. Centro, F. Ceradini, S. Cittolin, A. M. Cnops, C. Cochet, J. Colas, M. Corden, D. Dallman, S. D'Angelo, M. DeBeer, M. Della Negra, M. Demoulin, D. Denegri, R. Desalvo, D. DiBitonto, L. Dobrzynski, J. D. Dowell, M. Edwards, K. Eggert, E. Eisenhandler, N. Ellis, P. Erhard, H. Faissner, G. Fontaine, J. P. Fournier, R. Frey, R. Frühwirth, J. Garvey, S. Geer, C. Ghesquière, P. Ghez, K. L. Giboni, W. R. Gibson, Y. Giraud-Heraud, A. Givernaud, A. Gonidec, G. Grayer, P. Gutierrez, R. Haidan, T. Hansl-Kozanecka, W. J. Haynes, L. O. Hertzberger, C. Hodges, D. Hoffmann, H. Hoffmann, A. G. von Holtey, D. J. Holthuizen, R. J. Homer, A. Honma, W. Jank, P. I. P. Kalmus, V. Karimäki, R. Keeler, I. Kenyon, A. Kernan, R. Kinnunen, H. Kowalski, W. Kozanecki, D. Kryn, F. Lacava, J. P. Laugier, J. P. Lees, H. Lehmann, R. Leuchs, A. Leveque, D. Linglin, E. Locci, T. Markiewicz, G. Maurin, T. McMahon, J. P. Mendiburu, M. N. Minard, K. Morgan, M. Moricca, F. Muller, A. K. Nandi, L. Naumann, A. Norton, A. Orkin-Lecourtois, L. Paoluzi, G. Piano Mortari, M. Pimiä, A. Placci, M. Rabany, E. Radermacher, J. Ransdell, H. Reithler, J. Rich, M. Rijssenbeek, C. Roberts, C. Rubbia, B. Sadoulet, G. Sajot, G. Salvi, G. Salvini, J. Sass, J. Saudraix, A. Savoy-Navarro, D. Schinzel, W. Scott, T. P. Shah, M. Spiro, J. Strauss, K. Sumorok, F. Szoncso, C. Tao, J. Timmer, G. Thompson, E. Tscheslog, J. Tuominiemi, J. P. Vialle, G. Vismara, J. Vrana, V. Vuillemin, H. Wahl, P. Watkins, J. Wilson, M. Yvert, and E. Zurfluh, *Physics Letters B* 121 (1983): 77.

[8] TASSO Collaboration: R. Brandelik, W. Braunschweig, K. Gather, V. Kadansky, F. J. Kirschfink, K. Lübelsmeyer, H.-U. Martyn, G. Peise, J. Rimkus, H. G. Sander, D. Schmitz, A. Schultz von Dratzig, D. Trines, W. Wallraff, H. Boerner, H. M. Fischer, H. Hartmann, E. Hilger, W. Hillen, G. Knop, L. Koepke, H. Kolanoski, P. Leu, B. Löhr, R. Wedemeyer, N. Wermes, M. Wollstadt, H. Burkhardt, D. G. Cassel, D. Heyland, H. Hultschig, P. Joos, W. Koch, P. Koehler, U. Kötz, H. Kowalski, A. Ladage, D. Lüke, H. L. Lynch, P. Mättig, G. Mikenberg, D. Notz, J. Pyrlik, R. Riethmüller, M. Schliwa, P. Söding, B. H. Wiik, G. Wolf, R. Fohrmann, M. Holder, G. Poelz, O. Römer, R. Rüsch, P. Schmüser, I. Al-Agil, D. M. Binnie, P. J. Dornan, N. A. Downie, D. A. Garbutt, W. G. Jones, S. L. Lloyd, D. Pandoulas, J. Sedgbeer, R. A. Stern, S. Yarker, C. Youngman, R. J. Barlow, I. C. Brock, R. J. Cashmore, R. Devenish, P. Grossmann, J. Illingworth, M. Ogg, B. Roe, G. L. Salmon, T. R. Wyatt, K. W. Bell, B. Foster, J. C. Hart, J. Proudfoot, D. R. Quarrie, D. H. Saxon, P. L. Woodworth, E. Duchovni, Y. Eisenberg, U. Karshon, D. Revel, E. Ronat, A. Shapira, T. Barklow, J. Freeman, P. Lecomte, T. Meyer, G. Rudolph, E. Wicklund, S. L. Wu, and G. Zobernig, *Physics Letters B* 97 (1980):453.

[8] PLUTO Collaboration: C. Berger, H. Genzel, R. Grigull, W. Lackas, F. Raupach, A. Klovning, E. Lillestöl, J. A. Skard, H. Ackermann, J. Bürger, L. Criegee, H. C. Dehne, A. Eskreys, G. Franke, W. Gabriel, C. Gerke, G. Knies, E. Lehmann, H. D. Mertiens, U. Michelsen, K. H. Pape, H. D. Reich, M. Scarr, B. Stella, U. Timm, W. Wagner, P. Waloschek, G. G. Winter, W. Zimmermann, O. Achterberg, V. Blobel, L. Boesten, V. Hepp, H. Kapitza, B. Koppitz, B. Lewendel, W. Lührsen, R. van Staa, H. Spitzer, C. Y. Chang, R. G. Glasser, R. G. Kellogg, K. H. Lau, R. O. Polvado, B. Sechi-Zorn, A. Skuja, G. Welch, G. T. Zorn, A. Bäcker, F. Barreiro, S. Brandt, K. Derikum, C. Grupen, H. J. Meyer, B. Neumann, M. Rost, G. Zech, H. J. Daum, H. Meyer, O. Meyer, M. Rössler, and D. Schmidt, *Physics Letters B* 97 (1980):459.

[8] CELLO Collaboration: H.-J. Behrend, C. Chen, J. H. Field, U. Gümpel, V. Schröder, H. Sindt, W.-D. Apel, S. Banerjee, J. Bodenkamp, D. Chrobaczek, J. Engler, G. Flügge, D. C. Fries, W. Fues, G. Hopp, H. Küster, H. Müller, H. Randoll, G. Schmidt, H. Schneider, W. de Boer, G. Buschhorn, G. Grindhammer, P. Grosse-Wiesmann, B. Gunderson, C. Kiesling, R. Kotthaus, U. Kruse, H. Lierl, D. Lüers, T. Meyer, L. Moss, H. Oberlack, P. Schacht, M.-J. Schachter, A. Snyder, H. Steiner, G. Carnesecchi, A. Cordier, M. Davier, D. Fournier, J. F. Grivaz, J. Haissinski, V. Journé, A. Klarsfeld, F. Laplanche, F. Le Diberder, U. Mallik, J.-J. Veillet, A. Weitsch, R. George, M. Goldberg, B. Grossetête, F. Kapusta, F. Kovacs, G. London, L. Poggioli, M. Rivoal, R. Aleksan, J. Bouchez, G. Cozzika, Y. Ducors, A. Gaidot, S. Jadach, Y. Lavagne, J. Pamela, J. P. Pansart, and F. Pierre, *Physics Letters B* 110 (1982):329.

Complete References for Chapter 8

[6] G. Hanson, G. S. Abrams, A. M. Boyarski, M. Breidenbach, F. Bulos, W. Chinowsky, G. J. Feldman, C. E. Friedberg, D. Fryberger, G. Goldhaber, D. L. Hartill, B. Jean-Marie, J. A. Kadyk, R. R. Larsen, A. M. Litke, D. Lüke, B. A. Lulu, V. Lüth, H. L. Lynch, C. C. Morehouse, J. M. Paterson, M. L. Perl, F. M. Pierre, T. P. Pun, P. A. Rapidis, B. Richter, B. Sadoulet, R. F. Schwitters, W. Tanenbaum, G. H. Trilling, F. Vannucci, J. S. Whitaker, F. C. Winkelmann, and J. E. Wiss, *Physical Review Letters* 35 (1975):1609.

[11] S. R. Amendolia, G. Bellettini, P. L. Braccini, C. Bradaschia, R. Castaldi, V. Cavasinni, C. Cerri, T. Del Prete, L. Foa, P. Giromini, P. Laurelli, A. Menzione, L. Ristori, G. Sanguinetti, M. Valdata, G. Finocchiaro, P. Grannis, D. Green, R. Mustard, and R. Thun, *Physics Letters B* 44 (1973):119.

[12] U. Amaldi, R. Biancastelli, C. Bosio, G. Matthiae, J. V. Allaby, W. Bartel, G. Cocconi, A. N. Diddens, R. W. Dobinson, and A. M. Wetherell, *Physics Letters B* 44 (1973):112.

[16] V. Bartenev, R. A. Carrigan, Jr., I.-H. Chiang, R. L. Cool, K. Goulianos, D. Gross, A. Kuznetsov E. Malamud, A. C. Melissinos, B. Morozov, V. Nikitin, S. L. Olsen, Y. Pilipenko, V. Popov, R. Yamada, and L. Zolin, *Physical Review Letters* 31 (1973):1367.

[19] S. P. Denisov, S. V. Donskov, Yu. P. Gorin, A. I. Petrukhin, Yu. D. Prokoshkin, D. A. Stoyanova, J. V. Allaby, and G. Giacomelli, *Physics Letters B* 36 (1971):415.

[20] J. V. Allaby, A. N. Diddens, R. W. Dobinson, A. Klovning, J. Litt, L. S. Rochester, K. Schlüpmann, A. M. Wetherell, U. Amaldi, R. Biancastelli, C. Bosio, and G. Matthiae, *Physics Letters B* 34 (1971):431.

[21] G. Charlton, Y. Cho, M. Derrick, R. Engelmann, T. Fields, L. Hyman, K. Jaeger, U. Mehtani, B. Musgrave, Y. Oren, D. Rhines, P. Schreiner, H. Yuta, L. Voyvodic, R. Walker, J. Whitmore, H. B. Crawley, Z. Ming Ma, and R. G. Glasser, *Physical Review Letters* 29 (1972):515.
F. T. Dao, D. Gordon, J. Lach, E. Malamud, T. Meyer, R. Poster, and W. Slater, *Physical Review Letters* 29 (1972):1627.
J. W. Chapman, N. Green, B. P. Roe, A. A. Seidl, D. Sinclair, J. C. Vander Velde, C. M. Bromberg, D. Cohen, T. Ferbel, P. Slattery, S. Stone, and B. Werner, *Physical Review Letters* 29 (1972):1686.

[23] K. J. Foley, R. S. Jones, S. J. Lindenbaum, W. A. Love, S. Ozaki, E. D. Platner, C. A. Quarles, and E. H. Willen, *Physical Review Letters* 19 (1967):857.

[24] G. G. Beznogikh, A. Bujak, I. F. Kirillova, B. A. Morozov, V. A. Nikitin, P. V. Nomokonov, A. Sandacz, M. G. Shafranova, V. A. Sviridov, Truong Bien, V. I. Zayachki, N. K. Zhidkov, and L. S. Zolin, *Physics Letters B* 39 (1972):411.

[25] U. Amaldi, G. Cocconi, A. N. Diddens, R. W. Dobinson, J. Dorenbosch, W. Duinker, D. Gustavson, J. Meyer, K. Potter, A. M. Wetherell, A. Baroncelli, and C. Bosio, *Physics Letters B* 66 (1977):390.

[26] L. A. Fajardo, R. Majka, J. N. Marx, P. Némethy, L. Rosselet, J. Sandweiss, A. Schiz, A. J. Slaughter, C. Ankenbrandt, M. Atac, R. Brown, S. Ecklund, P. J. Gollon, J. Lach, J. MacLachlan, A. Roberts, and G. Shen, *Physical Review D* 24 (1981):46.

[27] M. Ambrosio, G. Anzivino, G. Barbarino, G. Carboni, V. Cavasinni, T. Del Prete, P. D. Grannis, D. Lloyd Owen, M. Morganti, G. Paternoster, S. Patricelli, and M. Valdata-Nappi, *Physics Letters B* 115 (1982):495.

[28] N. Amos, M. M. Block, G. J. Bobbink, M. Botje, J. Debaisieux, D. Favart, C. Leroy, F. Linde, P. Lipnik, J.-P. Matheys, D. H. Miller, K. Potter, C. Vander Velde-Wilquet, and S. Zucchelli, *Physics Letters B* 120 (1983):460.

[29] G. Barbiellini, M. Bozzo, P. Darriulat, G. Diambrini-Palazzi, G. De Zorzi, A. Fainberg, M. I. Ferrero, M. Holder, A. McFarland, G. Maderni, S. Orito, J. Pilcher, C. Rubbia, A. Santroni, G. Sette, A. Staude, P. Strolin, and K. Tittel, *Physics Letters B* 39 (1972):663.

A. Böhm, M. Bozzo, R. Ellis, H. Foeth, M. I. Ferrero, G. Maderni, B. Naroska, C. Rubbia, G. Sette, A. Staude, P. Strolin, and G. De Zorzi, *Physics Letters B* 49 (1974):491.

L. Baksay, L. Baum, A. Böhm, A. Derevshikov, G. De Zorzi, H. J. Giesen, H. Hilscher, J. Layter, P. McIntyre, F. Muller, B. Naroska, D. Reeder, L. Rossi, C. Rubbia, H. Rykaczewski, D. Schinzel, G. Sette, A. Staude, P. Strolin, G. J. Tarnopolsky, V. L. Telegdi, G. H. Trilling, and R. Voss, *Nuclear Physics B* 141 (1978):1.

[30] N. Kwak, E. Lohrmann, E. Nagy, R. S. Orr, W. Schmidt-Parzefall, K. R. Schubert, K. Winter, A. Brandt, F. W. Büsser, G. Flügge, F. Niebergall, P. E. Schumacher, J. J. Aubert, C. Broll, G. Coignet, H. de Kerret, J. Favier, L. Massonnet, M. Vivargent, W. Bartl, H. Dibon, H. Eichinger, C. Gottfried, G. Neuhofer, and M. Regler, *Physics Letters B* 58 (1975):233, 62 (1976):363, 68 (1977):374, *Nuclear Physics B* 150 (1979):221.

Subject Index

Name Index